HYPERSONIC VEHICLES
Emerging Concepts and Applications

HYPERSONIC VEHICLES
Emerging Concepts and Applications

by
Bhavna Shah | Surabhi Barman | Noor Khan

Wordpen Academics
Publishing Science That Matter

ISBN: 978-93-6084-079-2 (Hardback)

Published by : **WORDPEN ACADEMICS**
4736/23, Ansari Road, Darya Ganj,
New Delhi - 110 002
Phone: +91-011-35004336
E-mail: info@wordpenacademics.com

Printed at **:** **Replika Press Pvt. Ltd.**

Exclusive Distributor: ***Bio-Green Books, New Delhi***

Preface

"Hypersonic Vehicles: Emerging Concepts and Applications" takes readers on a thrilling journey into the forefront of aerospace engineering, exploring the cutting-edge technologies, innovative concepts, and transformative applications shaping the future of hypersonic flight. This book serves as a comprehensive guide for researchers, engineers, policymakers, and enthusiasts seeking to understand and harness the potential of hypersonic vehicles in aerospace exploration, defense, and transportation.

Hypersonic flight, defined as speeds exceeding Mach 5 (five times the speed of sound), represents the next frontier in aviation, offering unparalleled speed, range, and maneuverability compared to conventional aircraft. With the potential to revolutionize global transportation, enhance national security, and unlock new frontiers in space exploration, hypersonic vehicles hold immense promise for the future of aerospace technology.

As editors of this volume, we have endeavored to compile a diverse array of perspectives, insights, and breakthroughs from leading experts and innovators in the field of hypersonic flight. Through a blend of theoretical foundations, technological advancements, and practical applications, this book covers a wide range of topics, including hypersonic aerodynamics, propulsion systems, materials science, vehicle design, and operational challenges.

Our overarching goal in presenting this book is to inspire curiosity, spark innovation, and foster collaboration among stakeholders across academia, industry, and government. By providing a comprehensive overview of the state-of-the-art in hypersonic vehicles, we aim to empower readers to explore new horizons, push the boundaries of aerospace engineering, and contribute to the realization of hypersonic flight as a transformative technology for the 21st century.

We extend our sincere gratitude to all the contributors who have generously shared their expertise, experiences, and insights in the field of hypersonic vehicles. It is our hope that this book will serve as a catalyst for interdisciplinary dialogue, technological advancement, and visionary leadership in the pursuit of hypersonic flight and its myriad applications in the years to come.

Bhavna Shah | Surabhi Barman | Noor Khan

Contents

Section 1

Introduction

Chapter 1

Introductory Chapter: The Challenge to Fly Faster and Higher

1. Introduction

Starting from 1970s, commercial aviation has increased passengers' safety, comfort, and wellness. However, nothing has significantly changed about the length of the journeys since then.

A great improvement about travel speed was achieved by the Concorde and Tu-144 aircraft, when they started their commercial operations during the last two decades of the twentieth century, by flying at twice the speed of sound, exactly at $M_\infty = 2.04$ or about 2180 km/h.

But the supersonic era of commercial flights ended just in 2003 with the last flight of Concorde. Since then, attention to civilian supersonic flight has been lacking for about two decades. Currently, some US companies are showing renewed interest in supersonic transport for both business and common routes due to the availability of a higher technology readiness level (TRL) than in the past.

High TRLs and strategic plans, however, encouraged several research centers and industrials all over the world to look even further ahead toward the hypersonic flight.

Hypersonic flight is the capability to fly at very high speeds—more than Mach 5 or about 6000 km/h.

At hypersonic speed, we would fly over very long intercontinental routes with trip times of about 60 minutes. Therefore, the availability of a hypersonic aircraft will radically revolutionize the future of civilian transportation.

Potential advantages of hypersonic flight are numerous. Apart from reduced travel times, hypersonic flight will also improve access to space. In fact, hypersonic aircraft are expected to fly horizontally in Earth's atmosphere, like airplanes, and then proceed directly into orbit from a conventional runway, under the thrust of an air-breathing engine.

Today, much excitement and interest regard hypersonic vehicles. But, after about 60 years from the experimental hypersonic flights of the North American X-15 (see **Figure 1**), sustained hypervelocity travels are still an open issue of high-speed transportations [1].

Many aerospace agencies, large industries, and several start-ups are involved in many design activities and experimental campaigns both in wind tunnels and in-flight with full-scale experimental flying test beds and prototypes. Therefore, the dream of flying higher and faster with a hypersonic airplane, thus making hypersonic travel almost as easy and convenient as airliner travel, is increasingly becoming a reality.

Nowadays, flying at hypersonic speed is possible just to rocket launchers, in ascent flight, and unpowered re-entry vehicles both manned and unmanned or capsules entered another planet's atmosphere. However, some countries are going to plan or have claimed that have been already launched propelled hypersonic weapons even though for testing aims.

Figure 1.
US north American X-15. Courtesy of NASA.

2. Hypersonic operative scenarios

Generally speaking, hypersonic vehicles would operate within two scenarios, namely space applications and civilian applications.

2.1 Space applications

The space applications scenario belongs to space activities related to in-orbit activities and space exploration. Usually, hypersonic vehicles of this class are referred to as space planes and are unpowered and very blunt to slow down as much as possible while controlling the aerodynamic heating during re-entry or atmospheric entry (i.e. flying on other planets with atmosphere like Mars). So far, only three space planes have ever successfully flown: NASA's Space Shuttle, Boeing's X-37B, and the Soviet Buran, as shown in **Figure 2**.

Further, other countries are developing their own space plane like China and Europe. Few information about research developments of China is available due to reasons of confidentiality and secrecy. But, European Space Agency (ESA) is

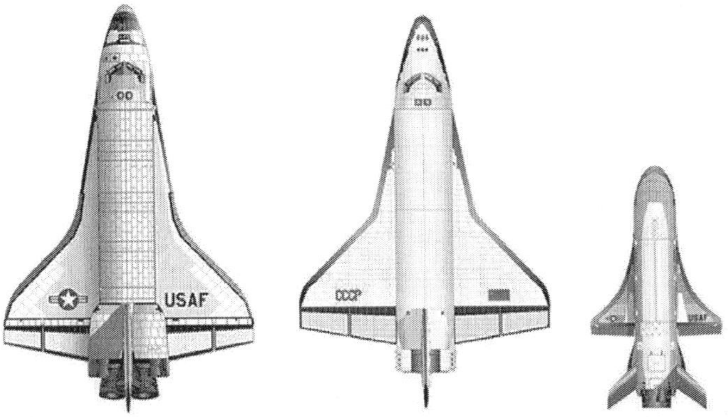

Figure 2.
US shuttle orbiter, soviet buran, and US X-37. Not to scale. Courtesy of Giuseppe De Chiara.

Figure 3.
Space rider. Courtesy of ESA.

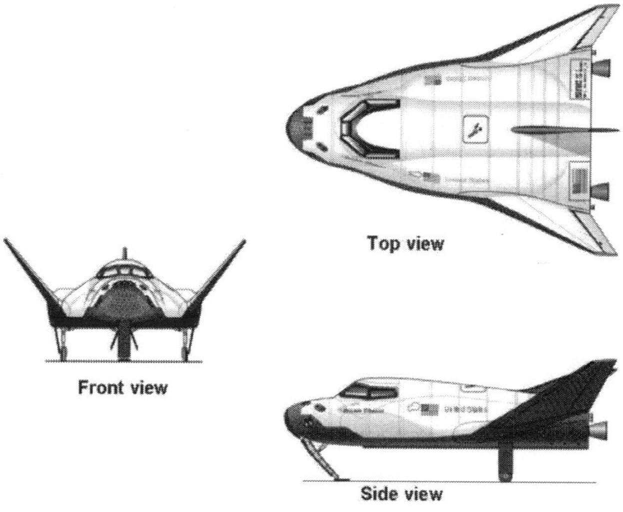

Figure 4.
SNC dream chaser. Courtesy of Giuseppe De Chiara.

expected to fly its own in-development space plane, namely Space Rider, whom maiden flight is expected in 2023, see **Figure 3** [2, 3].

More mature and nearly ready to flight is the two-stage space plane, namely Dream Chaser, under development by the Sierra Nevada Corporation (SNC) [4]. This space plane will support cargo resupply missions to and from the International Space Station (ISS), by 2022, and astronauts transfer by 2025, as well. With the Dream Chaser, the US will have again a commercial vehicle return from the ISS to a runway landing for the first time since the retirement of the NASA's space shuttle program (**Figure 4**).

2.2 Civilian applications

Civilian applications refer to slender, low drag aeroshapes that can enable sustained maneuvering flight in the atmosphere by exploiting air-breathing propulsion subsystems, like a scramjet or supersonic combustion ramjet engine. In fact, it is worth noting that the difference between supersonic and hypersonic flights is not a question of just giving more gas to the engines but requires different propulsive subsystems, namely ramjets. Fans which compress the air needed to fuel combustion in conventional turbofans engines would disintegrate at hypersonic speeds. Therefore, other types of thrusters without moving parts (i.e., ramjet) are needed.

Thus, rocket propelled aircraft carry their own reserve of liquid oxygen for combustion, thus giving them autonomy outside the atmosphere, but increase weight and volume.

On the contrary, scramjets provide thrust by exploiting atmospheric oxygen, which is compressed in the engine intakes to the aircraft's own speed, thus avoiding the use of turbines. This means that hypersonic aircraft must be characterized by very aggressive aerodynamic configurations, like waverider aeroshapes. A typical example of such a hypersonic configuration is provided in **Figure 5**, where the scramjet engine on the belly side of the aircraft is clearly visible [5].

A waverider-like configuration allows exploiting the inevitable formation of forebody shock wave (due to the high speed) to enhance its aerodynamic efficiency while properly feeding the air-breathing engine with incoming air through the aircraft bow shock, see **Figure 6**.

Figure 5.
Typical hypersonic aircraft [5].

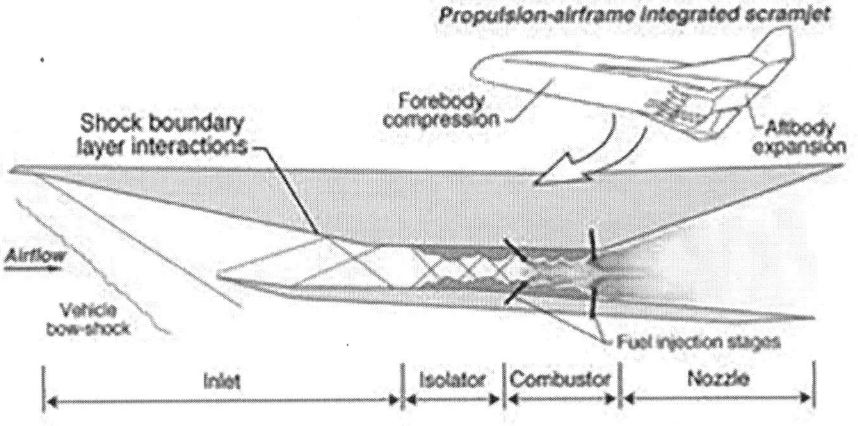

Figure 6.
Typical illustration to explain how air and fuel mix at supersonic speeds inside a scramjet. Courtesy of NASA [6].

This aircraft is expected to perform a variety of civilian (and military) missions, thus making hypersonic and access to space travels almost as easy and convenient as airliner travels. Therefore, the dream of flying higher and faster with a hypersonic airplane could become reality.

Anyway, the pioneering hypersonic flight was that of the North American X-15 even though it was achieved by exploiting a rocket engine. In November 1961, the X-15 flew at speeds over Mach 6, while on 3 October 1967, in California, an X-15 reached Mach 6.7. In the 1960s, this research program returned with valuable data that is still used in the development of spacecraft and aircraft today [1].

The first prove of a scramjet-powered flight was within the experimental X-43 scramjet program, namely Hyper-X Program, with a successful flight of the Boeing X-43A test bed in March 2004 [6]. The X-43 aircraft is shown in **Figure 7**.

During the experimental flight, the vehicle flew under its own scramjet power at an airspeed of Mach 6.8, or about 8046 km/h, for about 11 seconds. Then, on November 16, another scramjet-powered X-43A did it again, this time reaching hypersonic speeds above Mach 9.6, or about 10,943 km/h, in the final flight of the X-43A project. Both flights set world airspeed records for an aircraft powered by an air-breathing engine and proved that scramjet propulsion is a viable technology for powering future space-access vehicles and hypersonic aircraft.

Figure 7.
The X-43 aircraft. Courtesy of NASA [6].

Figure 8.
The X-51 aircraft. Courtesy of NASA [7].

Another successful experimental hypersonic vehicle, the Boeing X-51, underwent two successful tests from 2010 to2013. It completed its first unmanned power flight in May 2010, flying at a maximum speed of roughly Mach 5.1 (approximately) at 70,000 feet (**Figure 8**) [8].

An example of a research program on hypersonic aircraft carried out in Europe is LAPCAT (Long-Term Advanced Propulsion Concepts and Technologies) [9]. It is funded by the European Union and aimed at exploring the path to a hypersonic aircraft for transporting passengers at Mach 8 for antipodal flights, capable of traveling from Brussels to Sydney in 4 hours [6]. LAPCAT has been continued in other initiatives such as STRATOFLY (stratospheric flying opportunities for high-speed propulsion concepts), founded within the H2020 research and innovation program [10]. STRATOFLY investigates the feasibility of high-speed passenger stratospheric flight by taking into account for technological, environmental, and economic factors that allow the sustainability of new air space's exploitation, drastically reducing transfer time, emissions, and noise, and guaranteeing the required safety levels. In addition, STRATOFLY represents the first step toward future reusable launchers (**Figure 9**) [10].

Figure 9.
The LAPCAT-MR2 hypersonic cruiser concept [9].

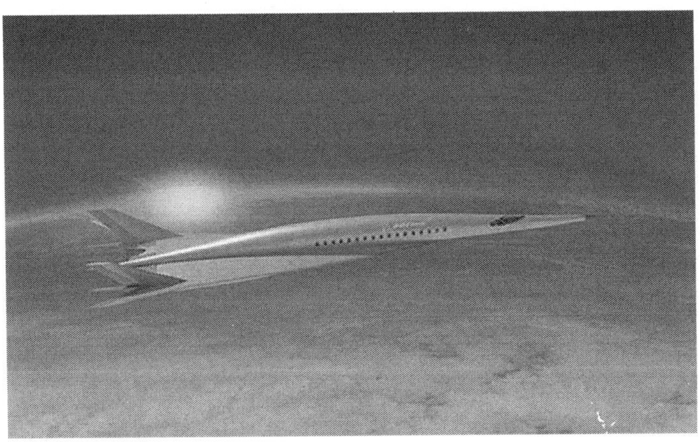

Figure 10.
Boeing M = 5 concept aircraft. Courtesy of Boeing [11].

Figure 11.
Hermeus M = 5 concept aircraft. Courtesy of Hermeus [8].

Another example of hypersonic aircraft is provided by Boeing in **Figure 10** [11]. It was presented in June 2018 as a concept for a Mach 5 aircraft, a speed chosen because it would allow the use of conventional materials such as titanium. At high speeds, in fact, the aircraft forebody can reach very high temperature due to aerodynamic heating, thus requiring the use of a ceramic heat shield.

Another promising hypersonic project is that of Hermeus company [8]. It is involved in the development of the Mach 5 aircraft, as shown in **Figure 11**.

3. Hypersonic flight and vehicle design

Accelerating an aircraft to hypersonic speeds demands huge efforts from both scientific and engineering points of view. Several design issues must be addressed including, for instance, multidisciplinary design optimization, advanced air-breathing propulsion, and thermal control. Therefore, flying at hypersonic speeds demands advanced vehicle design and testing, and creating critical technologies to overcome many related technical challenges, as well.

Several aerospace disciplines are involved in the hypersonic flight. Among others, we have avionics, aerodynamics, thermodynamics, materials technology, guidance and control systems, and propulsion.

One of the most important TRLs relies on the availability of high-temperature materials needed to shield the hypervelocity aircraft against the large aerodynamic heating the vehicle must withstand during flight. Friction with the atmosphere, in fact, exposes the aircraft aeroshape, and especially the aircraft forebody and leading edges, to extremely high temperatures of the order of 1000 K. Therefore, materials and strategies to cope with the high temperature, which threaten to melt and warp the structure, are fundamental.

Another important technology to make hypersonic flight reality is related to advanced propulsive subsystems. To counter the huge wave drag at high speed, reliable, effective, and efficient air-breathing propulsion systems are needed.

Then, also vehicle guidance, navigation, and control (GN&C) is fundamental provided that motion at hypersonic speed is extremely fast and the vehicle behavior in this flight regime must be accurately predicted.

As a result, the hostile environment hypersonic vehicles see in flight suggests that the challenge to routinely fly faster and higher for civilian transportation is too long to overcome.

To address these difficulties, a lot of research efforts are carried out by scientific and industrial international community to ensure that high-speed flight tests are successful. Some of these efforts are summarized in the present book, in terms of numerical and on-ground experimental activities.

Numerical tools, like computational fluid dynamics (CFD), prove to mathematically describe the hypersonic environment and predict the vehicle performance in flight, while on-ground facilities attempt to replicate the extreme environment in order to expose vehicles to the real expected flight conditions. But, even though CFD has made tremendous strides, it is not there yet as a viable replacement for real-world tests. Therefore, the more effective application of computational tools and experimental test facilities takes place when both are applied synergically on the same design analysis in a complementary fashion. For instance, the vehicle behavior is extensively investigated through a lot of cheaper numerical simulations, and the accuracy of numerical results is proved by means of a few expensive experimental test campaigns to compliment computer-based data collecting and to set realistic expectations on the technology.

This is a winning approach to progress toward the demonstration of readiness level of high-speed flight technologies.

Leveraging information and scientific investigation gather in the present book will deliver great benefit and help anyone involved in hypersonics to pursue his/her research aims both as a Ph.D. student and engineer.

References

[1] NASA Armstrong Fact Sheet: X-15 Hypersonic Research Program. Available from: https://www.nasa.gov/centers/armstrong/news/FactSheets/FS-052-DFRC.htmlNASA [Accessed: 02-02-2022]

[2] Available from: https://www.avio.com/it/space-rider [Accessed: 05-02-2022]

[3] Available from: https://www.asi.it/trasporto-spaziale/space-rider/ [Accessed: 06-02-2022]

[4] Available from: https://sierraspace.com/capabilities/dream-chaser-spaceplane/ [Accessed: 07-02-2022]

[5] Available from: http://www.aerospaceweb.org/design/waverider/ [Accessed: 10-02-2022]

[6] Available from: https://www.nasa.gov/centers/armstrong/news/FactSheets/FS-040-DFRC.html [Accessed: 11-02-2022]

[7] Available from: https://www.nasa.gov/topics/aeronautics/features/X-51A.html [Accessed: 11-02-2022]

[8] Available from: https://www.hermeus.com/ [Accessed: 15-02-2022]

[9] Available from: https://www.esa.int/Enabling_Support/Space_Engineering_Technology/Achievements_obtained_within_the_European_LAPCAT_program [Accessed: 12-02-2022]

[10] Available from: https://www.h2020-stratofly.eu/ [Accessed: 15-02-2022]

[11] Available from: https://www.boeing.com/features/2018/06/hypersonic-concept-vehicle.page [Accessed: 15-02-2022]

Section 2

Experimental Investigations

Chapter 2

Investigation of Hypersonic Conic Flows Generated by Magnetoplasma Light-Gas Gun Equipped with Laval Nozzle

Abstract

This chapter introduces new approach of hypersonic flow generation and experimental study of hypersonic flows over cones with half- angles $\tau 1 = 3°$ and $\tau 2 = 12°$. Mach number of the of the incident flow was M1 = 18. Visualization of the flow structure was made by the schlieren method. Straight Foucault knife was located in the focal plane of the receiving part of a shadow device. Registration of shadow patterns was carried out using high- speed camera Photron Fastcam (300 000 fps) with an exposure time of 1 μs. The Mach number on the cone was calculated from inclination angle of shock wave in the shadowgraph.

Keywords: hypersonic conic flows, light-gas magnetoplasma launcher, photometric shadow method

1. Introduction

Currently, light-gas cannons are one of the most effective devices for accelerating shells of relatively large mass to speeds comparable to the orbital. In this area of research, intensive and extensive experimental work is being carried out to study high-speed impact, resistance of materials to high-speed action of solid particles, hypersonic flows around axisymmetric bodies and are closely related to the development of aviation and space technology. In this case, some types of guns with the replacement of the accelerating channel by the Laval nozzle can be effectively used to study hypersonic flows around bodies, since the parameters of the working gas required for high-speed throwing and generation of a hypersonic air flow are identical. High-speed throwing system parameters and technical requirements are described in [1–5]. Since the 40s of the 20th century the generation of hypersonic flows, development of measurement methods and study of such flows structure over various test bodies were carried out by researchers from all over the world [6–16]. Numerical modeling methods mentioned, for example in [10, 11, 17–19], were added to the experimental tools of the research with the development of computer technology. However, it should be noted that experimental studies of hypersonic flows around the objects are still preferred. Due to the importance of problems to be solved flight tests are often used for higher authenticity of the experimental data. Usually total and static pressure is measured by means of Pitot-Prandtl probe to determine the Mach number, but in large number of research works measurements are not carried out due to complicated experimental techniques. Researchers [6, 7, 11–16] used the initial Mach number calculated from Laval nozzle geometrical parameters. Pressure measurement using a Pitot-Prandtl probe and temperature measurement using a thermocouple significantly distort the flow structure, complicate the design due to the need to place measuring elements inside and on the surface of the models under study, which, in the absence of flow visualization, can introduce uncontrollable errors in the experimental results. In hypersonic aerodynamic facilities preferable methods of research are optical methods with high-speed cameras as the reg-istering devices. Methods of optical visualization of supersonic flows include a method of a laser knife, smoke visualization, visualization by means of the spark discharge, etc. [13, 15, 16, 18]. Optical methods based on the relationship between the optical and thermodynamic properties of the medium (shadow methods, interference methods, spectral diagnostics) allow measurements in hypersonic flows with high accuracy. In some cases, to visualize and measure the temperature field on the surface of the model, a thermosensitive coating is used as an alternative to measurements using thermocouples [8, 9]. Research in the field of high-speed impact and diagnostics of hypersonic flow are often interrelated tasks, which makes it necessary to create experimental installations equipped with systems for optical visualization and diagnostics of the flow and measurement of the Mach number, which makes it possible to study hypersonic flow around objects without suppressing the investigated gas flow.

2. Experimental equipment for hypersonic flow generation and optical diagnostics

In most cases, to implement a hypersonic gas flow with a high Mach number, a shock tube equipped at the end with a diaphragm and a Laval nozzle is used [18–20]. In this case, the hot gas behind the front of the reflected shock wave flows out through the nozzle into the vacuum reservoir at a hypersonic speed. However, due to the large expansion of the gas, its density turns out to be rather low, which greatly complicates the visualization of the flow. Thus, it is possible to estimate the angles of deflection ε_x and ε_y of the probe light beam in geometrical optics approach while schlieren method of hypersonic flow visualization is used [21].

$$\varepsilon_x = \frac{1}{n_0}\int \frac{\partial n(x,y)}{\partial x} dz; \qquad \varepsilon_y \frac{1}{n_0}\int \frac{\partial n(x,y)}{\partial y} dz. \tag{1}$$

At the same time gradients of gas refraction index $\frac{\partial n(x,y)}{\partial y}$ and $\frac{\partial n(x,y)}{\partial x}$ are very small when the light gas outflows into vacuum. If helium is used as a light gas and the residual pressure in the vacuum chamber is 1 Torr and the characteristic scale of the streamlined body is $\sim 10^{-2}$ m, then the characteristic gradients of the refractive index do not exceed 5×10^{-6} m^{-1}. Under these conditions, the use of shadow methods and interference diagnosis becomes almost impossible. Therefore, in such cases, it is necessary to use the methods of multipass interferometry, multibeam interferometry, interferometry in polarized light, holographic methods or shadow methods with a large focal length and special imaging diaphragms [22]. On the other hand, the use of light-gas launcher where accelerating chan-nel is replaced with Laval nozzle allows to receive a hypersonic flow with high optical density and to apply Tepler method to its visualization. Hypersonic flow facility studied in this research is a modified light-gas launcher we used for ballistic tests and described in details in article [23]. The scheme of the experimental setup is shown in **Figure 1**.

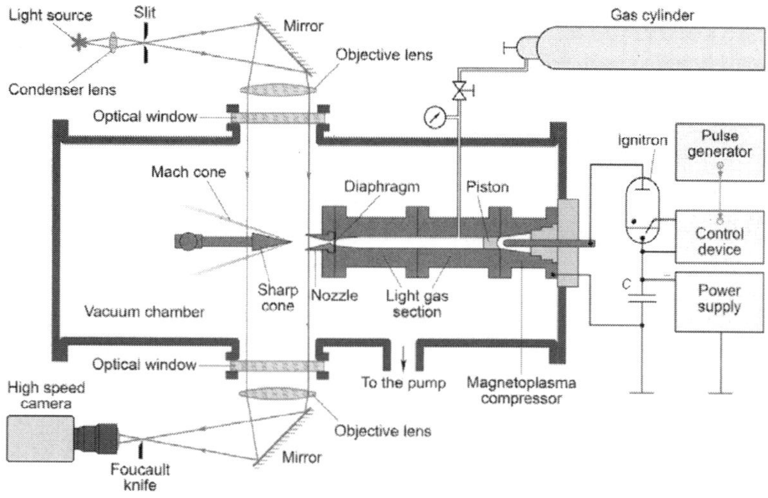

Figure 1.
Scheme of experimental facility.

This light-gas gun allows to accelerate balls with a diameter of 2.5 mm to 4 mm made of high-alloy steel to speeds of 2.5–4 km/s.

During the experiments, the light gas section was filled with helium to a pressure of 40 bar. Experiments on hypersonic flow around cones were carried out in a vacuum chamber with a residual gas pressure of 1 Torr. As a result of a high-current discharge in a magnetoplasma accelerator filled with gunpowder, the piston is set in motion and causes an adiabatic compression of the light gas. The capacitor bank of 1 200 μF was charged to the voltage 4.5 kV. In front of the confuser of Laval nozzle 5 brass diaphragms 100 μm thick each were placed. The rupture of diaphragms occurred when pressure in light-gas section reaches ~1 600 bars. Compression rate and temperature of working gas can be calculated from the Poisson equation:

$$\frac{T_2}{T_1} = \left(\frac{V_1}{V_2}\right)^{\gamma-1} = \left(\frac{P_2}{P_1}\right)^{\gamma-1/\gamma}, \tag{2}$$

where T is gas temperature, V is gas volume, P is gas pressure and $\gamma = \frac{5}{3}$ for the monoatomic gas. Subscript 1 corresponds to initial state of gas and subscript 2 is for state of gas at diaphragms rupture. When diaphragms rupture compression rate of the working gas is about 10 and temperature is about 1500 K. With the simultaneous outflow of gas through the critical section of the Laval nozzle, further compression of the gas continues until the piston reaches the end point. The maximum compression ratio is about 50 and can be calculated as the ratio of the volume of the light gas section to the volume of the confuser of the Laval nozzle. The ratio of the throat to the outlet of the diffuser in the Laval nozzle for a Mach number of 18 was calculated using the formula [22, 24]:

$$\frac{A}{A*} = \left(\frac{2}{\gamma+1}\right)^{\frac{\gamma+1}{2(\gamma-1)}} \frac{1}{M} \left(1 + \frac{\gamma-1}{2} M^2\right)^{\frac{\gamma+1}{2(\gamma-1)}}, \tag{3}$$

where A is area of Laval nozzle exit section and $A*$ is area of Laval nozzle throat. The test cones were fixed coaxially with the nozzle at a distance of 20 mm from the nozzle outlet. To visualize the hypersonic gas flow, the shadow knife-and-slit method was used. Width of a slit was 0.16 mm. Foucault's knife was placed in the focal plane of a receiving part of the shadow device and shaded a half of the image of a slit. The 150 W halogen lamp was used as a light source. The focal length of the shadow device was 1 000 mm. Registration of shadow patterns was carried out using high-speed camera Photron Fastcam (300 000 fps) with an exposure time of 1 μs (**Figure 2**).

3. Experimental results and discussion

The analysis of shadow pictures sequence shows that at initial time the flow contains some particles formed as a result of diaphragm destruction and about each of which Mach cone (**Figure 2**) is observed. Further there is an increase of speed of a flow during 700 μs and its subsequent recession during 3 400 μs. At the maximum speed of the flow in shadow pictures the most acute angle of a shock wave inclination is observed. In **Figures 3** and **4** shadow pictures of a hypersonic flow over

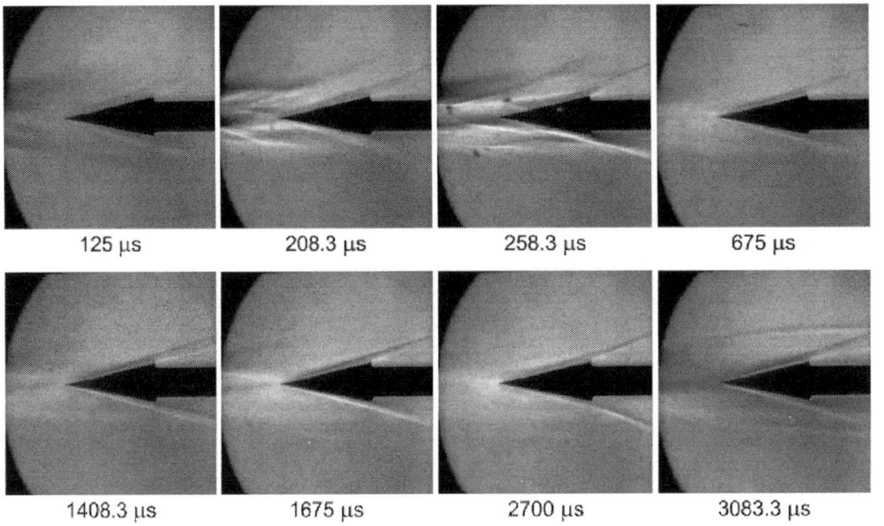

Figure 2.
Shadowgraph images of a hypersonic flow over cone with half-angle $\tau_2 = 12°$.

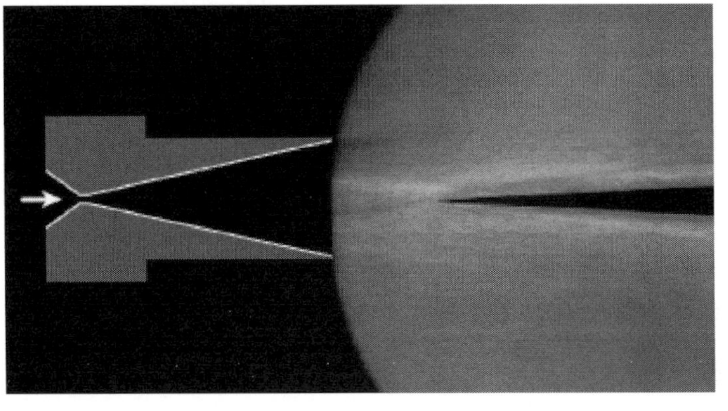

Figure 3.
Shadowgraph of hypersonic flow over the cone with a half-angle $\tau_1 = 3°$.

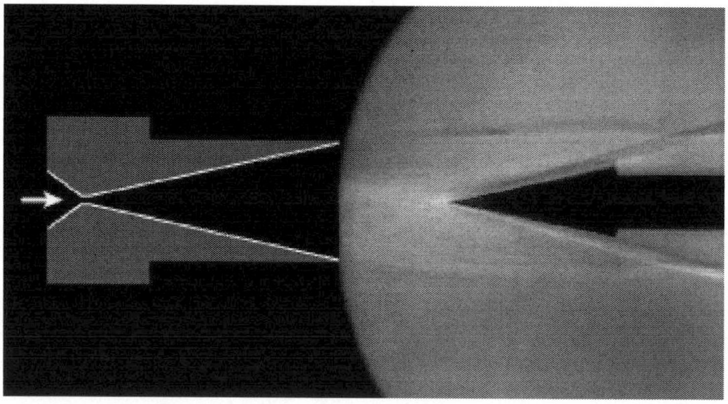

Figure 4.
Shadowgraph of hypersonic flow over the cone with a half-angle $\tau_2 = 12°$.

cones with half angles $\tau_1 = 3°$ and $\tau_2 = 12°$ are presented at maximum speed of the hypersonic flow.

Correlation of Mach number and shock wave inclination angle observed on shadow images of hypersonic flow over sharp cone was described in details by [25] for the first time. Calculation of Mach number was carried out on the basis of measurement of shock wave inclination angle σ observed on shadow pictures (**Figures 3** and **4**) using formula [26]:

$$M_1 = \csc\sigma \sqrt{\frac{2(B + C - \sec)}{(\gamma+1)\left(B + C + \frac{\cos\sigma}{\sin^2\sigma}\right) - (\gamma-1)(B + C - \sec\sigma)}} \tag{4}$$

where

$$B = -\ln\left(\frac{\sin\tau}{1-\cos\tau}\right) - \frac{\cos\tau}{\sin^2\tau}; \qquad C = \ln\left(\frac{\sin\sigma}{1-\cos\sigma}\right). \tag{5}$$

The Mach number calculation results were M1 = 18 for hypersonic flow over cone with half-angle $\tau_1 = 3°$ and $M_2 = 14.4$ for the cone with $\tau_2 = 12°$.

4. The analysis of shadowgraphs and calculation of helium density field at a hypersonic flow over a cone

As it is known from the theory of the Schlieren technique [27], relative change of light intensity $\Delta I / I_0$ in geometrical optics approach in case of small deflection angles of the light ε is directly proportional to the magnitude of the deflection angle of the light rays in the optical inhomogeneity and can be defined by a ratio

$$\frac{\Delta I}{I_0} = \frac{\varepsilon f}{d}; \qquad \Delta I = I - I_0, \tag{6}$$

where d = 0:16 mm is a width of a slit, f = 1 000 mm is a focal length of optical system, I_0 is light intensity in the absence of an optical inhomogeneity, I is light intensity in the presence of an optical inhomogeneity. Since the streamlined body has an axial symmetry, when processing shadow pictures the assumption of axisymmetric structure of the flow was used. The coordinate system is chosen in such a way that the x-axis is perpendicular to the edge of the imaging knife, the z-axis is directed along the probing light beam of the shadow device, and the y-axis is along the symmetry axis of the cone and the flowing stream. The origin is set at the apex of the cone. In this coordinate system, the Abel equation can be written in the form (where R is the radius of the inhomogeneity) [21, 27, 28]:

$$\varepsilon(x) = 2\int_x^R \frac{d\ln n(r)}{dr}\frac{xdr}{\sqrt{r^2-x^2}}. \tag{7}$$

For calculations we use Abel equation transformation where $d \ln n/dr$ is expressed explicitly:

$$\frac{d\ln n}{dr} = -\frac{1}{\pi}\frac{d}{dr}\int_r^R \frac{\varepsilon(x)dx}{\sqrt{r^2-x^2}}. \tag{8}$$

After repeated integration [21, 27, 28] the distribution of refraction index is received:

$$\ln\frac{n(r)}{n_0} \approx \frac{n(r)-n_0}{n_0} = \frac{\Delta n}{n_0} = \frac{1}{\pi}\int_r^R \frac{\varepsilon(x)dx}{\sqrt{x^2-r^2}}, \tag{9}$$

where n_0 is known in advance value of refraction index on the line of integration, for example in the region of undisturbed flow $n_0 = n(R) \cong 1$.

The flow area was divided into annular zones, the number of which was selected in accordance with the resolution of the video camera matrix. The deflection angle within the annular zone $[r_i; r_{i+1}]$ is assumed to be constant. Then the expression (9) can be written as a sum of elementary integrals:

$$\frac{\Delta n(r_j)}{n_0} = \frac{1}{\pi}\sum_{i=j}^{N-1}\varepsilon(r_i)\int_{r_i}^{r_{i+1}} \frac{dr}{\sqrt{r^2-r_j^2}}. \tag{10}$$

The distribution of the refractive index in the flow is calculated as a result of independent integration at each section separately. With such processing of shadow, characteristic distortions arise in the distribution of the refractive index, which are associated with a high level of noise in the experimentally measured light intensity along the y axis. To eliminate random noise, a preliminary smoothing of the luminous intensity function along the y-axis was carried out using the least squares method [29–32]:

$$\int_{y_{min}}^{y_{max}} I''\left(y^2\right)dy \to min. \tag{11}$$

At the same time the smoothed function values should differ from the experimental function values not more than the standard deviation δ of the probe light intensity I_0 noise. Calculation of helium density ρ was carried out with use of Gladstone-Dale Equation [21, 28].

$$n - 1 = K_{He}\rho, \tag{12}$$

where $K_{He} = 0.19607\,\text{cm}/\text{g}^3$ is Gladstone-Dale constant for helium. The value for the given gas and given wavelength is considered to be a constant in the wide

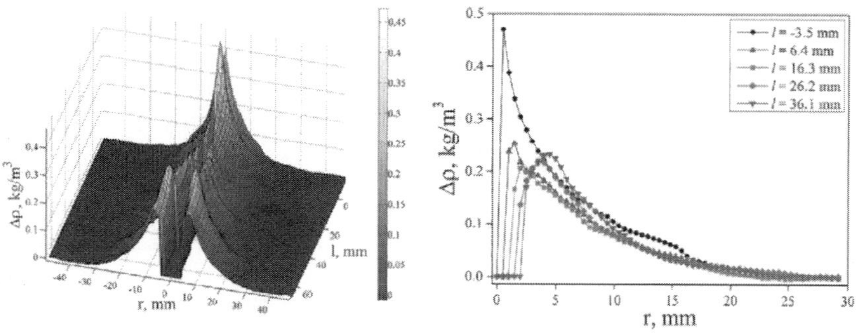

Figure 5.
Results of density field calculations (hypersonic helium flow past cone with the half-angle $\tau_1 = 3°$).

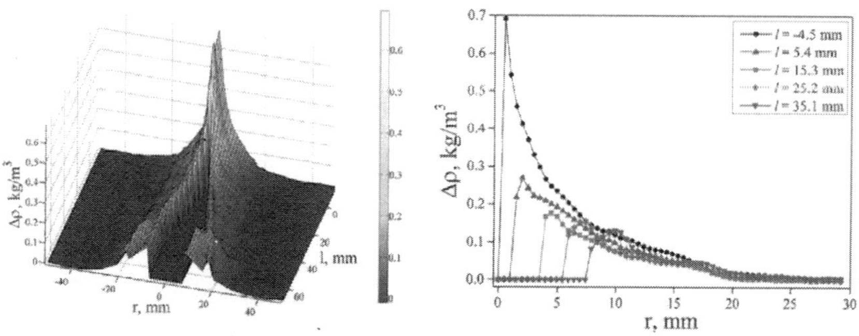

Figure 6.
Results of density field calculations (hypersonic helium flow past cone with the half-angle $\tau_2 = 12°$)

pressure range. Its value was calculated for gas under normal conditions with use of tabular values of density and refraction index [21]. As far as the initial values n_0 and ρ_0 are arbitrarily chosen values, we can calculate $\Delta\rho$ from the following equation:

$$\Delta n = K_{He} \Delta\rho. \tag{13}$$

Calculation results for helium density fields at a hypersonic flow over the cones with a half-angle $\tau_1 = 3°$ and $\tau_2 = 12°$ are shown in **Figures 5** and **6**.

5. Conclusions

The use of light gas ballistic systems, in which the projectile acceleration channel is replaced by a Laval nozzle, makes it possible to generate hypersonic flows with a high Mach number. The gas density of such flows is sufficient for photometric optical measurements by shadow or interference methods. This makes it possible to study hypersonic flow models of any complexity. In addition, due to the high density of the hypersonic flow, it becomes possible to experimentally study the destructive effect of the hypersonic gas flow on the bows of hypersonic missiles and aircraft.

References

[1] BULAH, B. M. 1970 Nonlinear Conic Gas Flows. *Moscow Nauka* **[In Russian]** 343p.

[2] GOLUBYATNIKOV, A. N. , PILYUGIN, N. N. & LEONTIEV, N. E. 2003 Methods of Upgrading the Efficiency of Light-gas Guns. *Usp. Meh.* vol. 2 (**1**), pp. 97–124.

[3] SMITH, F. 1963 Theory of a Two-Stage Hypervelocity Launcher to Give Constant Driving Pressure at the Model. *J. Fluid Mech.* vol. 17 (**1**), pp. 113–125.

[4] PUTZAR, R. & SCHAEFER, F. 2015 EMI's TwinGun Concept for a New Light-Gas Gun Type Hypervelocity Accelerator. Procedia Eng. , vol. 103, pp. 421 – 426.

[5] PUTZAR, R. & SCHAEFER, F. 2016 Concept for a New Light-Gas Gun Type Hypervelocity Accelerator. Int. J. Impact Eng. **88**, 118 – 124.

[6] ABRAMOVICH, G.N. 1973 Applied Gas Dynamics. Defense Technical Information Center. ed. 3rd.

[7] ROGERS, E. W. E., BERRYAND, C. J. & Davis, B. M. 1971 Experiments with Cones in Low-Density Flows at Mach Numbers near 2. Her Majestys Stationery Office **3505**.

[8] HUBNER, J. P. & CARROLL, B. F. & SCHANZE, K. S. & JI, H. F. & HOLDEN, M. S. 2001 Temperature- and Pressure-Sensitive Paint Measurements in Short-Duration Hypersonic Flow. *AIAA J./* vol. 39(**4**), 564–659.

[9] KURITA, M. & NAKAKITA, K. & MITSUO, K. & WATANABE, S. 2006 Temperature Correction of Pressure-Sensitive Paint for Industrial Wind Tunnel Testing. *J. Aircr./* vol. 43(**5**), 1499–1505.

[10] KUSSOY, M. I. & BROWN, J. D. & BROWN, J. L. & LOCKMAN, W. K. & HORSTMAN, C. C. 1988 Fluctuations and Massive Separation in Three-Dimentional Shock-Wave/Boundary-Layer Interactions. *Tech. Rep.* 89224. NASA Tech. Mem.

[11] BROWN, J. D.& BROWN, J. L. & KUSSOY, M. I. 1988 Fluctuations and Massive Separation in Three-dimentional Shock-Wave/Boundary-Layer Interactions. *Tech. Rep.* 101008. NASA Tech. Mem.

[12] HORSTMANAND, C. C. & KUSSOY, M. I. 1992 Intersecting Shock-Wave/ Turbulent Boundary-Layer Interactions at Mach 8.3. *Tech. Rep.* 103909. NASA Tech. Mem.

[13] RADZIG, A. N. 2004 Experimental Hidroaeromechanics. *Moscow MAI* **[In Russian]**.

[14] LEUTIN, P. G. 1976 Preasure Distribution on Sharp Cones at Angle of Sttack $\alpha \approx 10°$ in Supersonic Flow. *TsAGI Sci. J.* vol. 7 (**2**), 163–166.

[15] RIDYARD, H. W. 1954 The Aerodynamic Characteristics of Two Series of Lifting Bodies at Mach Number 6.86. *Tech. Rep.* L54C15. NASA Tech. Mem.

[16] WARREN, W. R. & KAEGI, E. M. & HARRIS, C. J. & GEIGER, R. E. 1962 Shock Tunnel Studies of the Aerodynamics of Atmospheric Entry *Tech. Rep.* R62SD56. General Electric CO.

[17] ZHIVOTNOV, S. D. & NIKOLAEV, V. S. & PROVOTVOROV, V. P. 1999 Aerodynamic Characteristics of Cones at Attack Angle Under Viscid-Nonviscid Interaction at Supersonic Velocities. *TsAGI Sci. J.* vol. 30 (**3–4**), 97–105.

[18] ERDEM, E. & YANG, L. & KONTIS, K. 2009 Drag Reduction by Energy Deposition in Hypersonic Flows. In *16th AIAA/DLR/DGLR Int. Space Planes Hypersonic Syst. Technol. Conf.* 19-22 Oct., Bremen, Germany. *AIAA Paper 2009-5347.*

[19] Saravanan, S. & Jagadeesh, G. & Reddy, K. P. J. 2009 Convective Heat-Transfer Rate Distributions Over a Missile Shaped Body Flying at Hypersonic Speeds. *Exp. Therm. Fluid Sci*. vol. 33 (**4**), 782–790.

[20] Gaydon, A. G. & Hurle, I. R. 1963 The Shock Tube in High Temperature Chemical Physics. Verlag Chapman and Hall Ltd.

[21] Vasilev, L. A. 1966 Shadow Methods. [**In Russian**] Nauka.

[22] Belozerov, A. F. 2007 Optical Methods of Gas Flow Visualization. [**In Russian**] Kazan State Techn. Univ.

[23] Khramtsov, P. P. & Vasetskiy, v. A. & Grishchenko, V. M. & Makhnach, A. I. and Chernik, M. Yu. and Shikh, I. A. 2015 Experimental and Analytical Study of a Two Stage Light-Gas Magnetoplasma Launcher for Ballistic Tests under Vacuum Conditions. *High Temp. Mater. Processes* vol. 19 (**3–4**), 209–219.

[24] Anderson, J. D. 2001 Fundamentals of Aerodynamics. McGraw-Hill.

[25] Liepmann, H. W. & Roshko, A. 1957 Elements of Gas Dynamics. John Wiley & Sons Inc.

[26] Pottsepp, L. 1960 Inviscid Hypersonic Flow Over Unyawed Circular Cones. *J. Aerosp. Sci*. vol. 27 (7), 558–559.

[27] Taylor, H. S. and Ladenburg, R. W. 1955 Physical Measurements in Gas Dynamics and Combustion. Oxford University Press.

[28] Skotnikov, M. M. 1976 Qualitative Shadow Methods in Gas Dynamics. [**In Russian**] Nauka.

[29] Savitzky, A. and Golay, M. J. E. 1964 Smoothing and Differentiation of Data by Simplified Least Squares Procedures. *Anal. Chem*. vol. 36 (**8**), 1627–1639.

[30] Steinier, J. & Termonia, Y. and Deltour, J. 1972 Comments on Smoothing and Differentiation of Data by Simplified Least Square Procedure. *Anal. Chem*. vol. 44 (**11**), 1906-1909.

[31] Bockasten, K. 1961 Transformation of Observed Radiances into Radial Distribution of the Emission of a Plasma. *J. Opt. Soc. Am*. vol. 51 (**9**), 943–947.

[32] Gander, W. and Hrebicek, J. 2004 Solving Problems in Scientific Computing Using Maple and MATLAB. Springer-Verlag Berlin Heidelberg.

Chapter 3

Experimental Analysis of Waverider Lift-to-Drag Ratio Measurements in Rarefied and Supersonic Regime

Abstract

This work, performed in the MARHY rarefied hypersonic facility, experimentally explores the effects of rarefaction on a classical waverider geometry. This hypersonic vehicle is designed to develop a shock attached along the leading edge length to improve flight efficiency. The concept was first proposed by Nonweiler in 1959. Since then, many studies have been conducted, mainly on numerical aspects. Few works have included the influence of the viscous effect, we can cite those of Bowcutt who showed how viscous effects impact the optimal shapes due to the skin friction drag. However, the trajectories of these types of vehicles anticipate flights with high Mach numbers and at high altitudes where rarefaction effects can strongly impact the lift-to-drag ratio predictions. This work focuses on the behavior of the L/D ratio at different supersonic operating conditions. The viscous effects were analyzed with 4 operating flow conditions: Mach 2 and 8 Pa static pressure and Mach 4 with 2, 8 and 71 Pa static pressures. For this purpose, the aerodynamic coefficients were measured for several angles of incidence. with a homemade sting balance. The experimental results were compared to Monte Carlo numerical simulations performed with the DS3V code.

Keywords: Waverider, Supersonic rarefied regime, Wind tunnel, Lift-to-Drag ratio, Sting balance, DSMC

1. Introduction

The classic hypersonic waverider is a vehicle geometry designed to capture the post-shock flow field between the waverider body, optimized at a specific Mach number and the flow. This specific geometry produces a high pressure region at the bottom surface of the vehicle that maximizes the lift-to-drag ratio (L/D). In general, hypersonic glider is designed to operate at Mach numbers higher than 5. This particular class of vehicle provides a higher L/D ratio than a generic vehicle for the same angle of incidence. However, the real advantage of hypersonic gliders lies in the fact that, for the same lift, the angle of incidence of the glider is much lower, which implies a low pressure drag compared to the generic vehicle. A very large number of geometrical configurations of hypersonic gliders have been developed since the late 1950s. One can refer to the recent review article published by

Ding et al. [1] which gives an overview of the different hypersonic glider geometries published to date. However, the usual methodology to create a hypersonic glider geometry remains the same as it is based on the generation of a shock from a canonical geometry: cone, von Karman warhead, wedge, etc. Depending on the desired characteristics (Mach, altitude, angle of incidence, etc.), the shape of the hypersonic glider is generated by projection onto the shape of the shock wave generated by the canonical shape. This implies that a hypersonic glider is only defined for a given operational configuration, nevertheless very recent works have generated geometries optimized for several Mach numbers and tested by numerical simulations (Mach 5 and 10) [2].

When a hypersonic glider evolves at high altitude in a rarefied flow, additional flow characteristics must be taken into account for the design of glider geometries. Indeed, viscous effects appear, it is possible to take them into account in the initial definition of the geometry [3, 4]. In particular, the process of viscous interactions (strong and/or weak) will generate a modification of the shape of the shock waves: in the case of a strong interaction, the boundary layer that develops will "push" the shock outwards. The fluid layer between the shock and the boundary layer will then act on the development of the boundary layer, thus setting up a phenomenon of mutual interactions [5]. In addition, as the flow becomes thinner, especially near the leading edges, the shock waves become more and more detached from the geometry, leading to interactions between the lower and upper parts of the glider on both sides. This results in a significant decrease in lift due to the "emptying" (spillage) of the high pressures initially located under the glider and which are the origin of the lift effect sought for hypersonic gliders. Moreover, the progressive development of viscous drag, especially on the upper part of the geometry, inevitably leads to an increase in the overall drag [6]. The combination of these two effects leads to a strong decrease of the L/D ratio, thus strongly degrading the aerodynamic performance of hypersonic gliders. These effects have been demonstrated by numerical simulations but, to date, there are no experimental results in the literature that have demonstrated and quantified them.

It can also be noted that for Mach numbers of about 15, rarefaction effects can begin to appear from 40 km altitude. As a reminder, it is generally considered that we are in the presence of a rarefaction flow for altitudes above 60 km. This result was indirectly confirmed by Rault [7] who showed, via numerical simulations, that the flow in the vicinity of a hypersonic glider was mainly in rarefaction regime whereas the Knudsen number calculated with infinite flow conditions indicated rather that the flow was in continuous regime. Thus, rarefaction effects may be present at a much lower altitude than expected.

To take into account these rarefaction effects, a new category of hypersonic gliders was created in the 1980s with the pioneering work of Professor Anderson of the University of Maryland [8–10], introducing the "viscous optimized waveriders". Currently, most of this type of waverider is studied in China, in particular Liu et al. [11] shows that the considered altitude (thus the level of rarefaction) influences significantly the shape of the gliders, playing notably on the volume of the payload.

Regardind the identification of rarefaction effects and their quantification, to our knowledge only results from numerical simulations exist [12–14]. Experimental results from wind tunnel tests are currently absent from the literature. Indeed, a quick overview of the experimental work undertaken on hypersonic gliders shows the lack of experimental data concerning rarefied flows in the slip or transition regime. This is explained by the lack of experimental facilities to simulate this flow regime at high Mach number, but also by the metrological difficulties inherent in the characterization of rarefied flows. Concerning hypersonic gliders, the studies presented in the literature only concern medium altitudes: 22.5 km [15] or 30.6 km [16]. The very few studies conducted in the rarefied regime [17–19], contain only fragmentary results which do not allow to highlight the effects of rarefaction on the aerodynamic performance of hypersonic gliders, and even less to quantify them over a wide range of flow conditions.

To overcome the lack of experimental data available in the literature, correlation functions are often used to extrapolate the aerodynamic behavior of these types of vehicles at high altitude. These correlations are based on old experimental data or are relative to canonical (thus simplified) geometries [20].

Moreover, these semi-empirical laws often use parameters based on quantities that must be measured locally, such as pressure or wall temperature [21]. Thus, the use of existing bridging functions can lead to an approximate evaluation of the aerodynamic coefficients of hypersonic gliders, especially since the diversity of existing geometries is large. It is therefore necessary to establish empirical correlations suitable for hypersonic gliders in rarefied flow at high Mach number.

By exploring a field that is currently poorly studied experimentally, the knowledge acquired on rarefied hypersonic flows around complex 3D geometries will be directly useful to actors planning the development of new atmospheric re-entry vehicles, such as space agencies and aerospace industries. In particular, the hypersonic glider concept is also being considered as an atmospheric re-entry or planetary transfer vehicle: [22, 23] for space exploration of telluric planets, gas giants, or some of their moons, for which trajectory control is also central to the success of a space exploration mission [24].

The main objective of this study is to determine experimentally if rarefaction effects significantly modify or not the aerodynamics of hypersonic gliders when they fly at high altitude in rarefied atmosphere. This study will be carried out with the MARHy rarefied hypersonic wind tunnel of the FAST experimental platform of the ICARE laboratory, which allows to cover the flight conditions in terms of Reynolds/Mach of glider trajectories for altitudes ranging from 100 km to 60 km. This paper focuses on measurements of the aerodynamic forces of a classical hypersonic glider in Mach 4 and Mach 2 flows with different Knudsen numbers in order to study the correlations between viscous effects and L/D ratios.

2. The model definition

2.1 General waverider concepts

Lifting vehicles founded a real interest since the late 1960s, in USA with the development of the Space Shuttle orbiter design based on a flat delta plate shape. In the same time, much works originated from Europe focused on more generalized research which differ from the pattern of American research. In particular, European effort concerned the aerodynamics of lifting reentry in hypersonic regime with the design and performance of so called waveriders [25]. The concept of waverider was introduced in 1959 by Nonweiler [26] as a type of delta wing and was named caret waverider. During its works, Nonweiler noted that the shock create a high pressure on intrados (the under surface) and induced lift. The main concept to design a waverider can be summarized by the idea that the lower surface is generated from the streamlines of the flow over a body and the upper surface is aligned with the free stream flow. During the early 1980s, waveriders design have been improved thanks to Rasmussen studies [27]. Rasmussen used the shock created by the flowfield in contact with a canonical geometry to generate lift. This method allowed to obtain waveriders with better performances than Nonweiler's design.

Based on Ramussen method other studies were carried out to optimize the geometry and increase the Lift-to-Drag ratio, giving birth to various families of cone derived geometries as well as hybrid variations like cone-wedge waveriders [28, 29]. Waveriders are designed to maximize lift-to-drag for a flight specific conditions (Mach, altitude). Nevertheless during the flight, altitude and speed will change getting out of their optimized flight domain range leading to uncertainties regarding their aerodynamic behavior [30, 31].

2.2 The experimental waverider definition

This experimental work intends to study the influence of rarefaction effects on the aerodynamic performances of the waverider (lift, drag and L/D ratio). To do so, we have looked for an optimal geometry close to our experimental test conditions in rarefied regime. The geometry we have designed is based on the Rolim geometry which is optimized for an altitude between 40 km and 50 km and a speed of Mach 10. This geometry, if not optimal, has been optimized for altitudes close to the one simulated in our wind tunnel between 50 km and 80 km. Concerning the Mach number, the experimental tests have been carried out at Mach 2 and Mach 4, far from the Mach 10 conditions proposed by Rolim. This Mach number discrepancy allowed to study the aerodynamic behavior of the waverider for speeds lower than the expected one as well as its evolution outside its privileged flight domain. The waverider is 100 mm long, 35.7 mm wide, 6.6 mm high, **Figure 1**, and the dihedral angle, Φ, is approximately 53°, and it was 3D printed in resin. The dimensions of the waverider were chosen according to the test area of the wind tunnel.

3. Aerodynamic forces measurements

3.1 The sting balance description

An aerodynamic balance has been developed to measure drag and lift forces in the MARHy operating flow conditions. The main difficulty comes from the fact that operating conditions concerns rarefied regime with low density flows, giving force values estimated between 1mN and 1 N. Aerodynamic balances applied to supersonic wind tunnels are ranged in two main balance groups' internal and external balances. Internal balances [32] are a test model extension and located inside the test section, on the contrary, external [33] are placed outside the test section and the model is supported with thin wires. For the present study, a sting balance was designed to measure drag and lift forces applied to the waveriders in supersonic and hypersonic flows. The design of this device had to meet to several constraints: be able to measure small forces values around mN, to have an aerodynamic design to not obstruct the test section and ability to measure aerodynamic model forces with different incidence angles without changing the balance calibration. This sting balance is composed in two parts, drag and lift modules as showed on **Figure 2**. The principle adopted for force measurement is based on the deformation of thin slats equipped with strain gauges, deforming when the test model is an incoming flow.

The drag module as can be seen on **Figure 3** is designed in three parts: upper part, slats, and lower part in order to facilitate manufacturing. There are six slats arranged in three rows and two columns. Indeed, many configurations have been dynamically simulated with a CAO software to optimized de number of lamelles, their arrangement and their thickness. The **Figure 3** shows the deformation of the drag module slats when subjected to a horizontal force. The results showed that this configuration is the best compromise to optimize the bending, and to minimize the

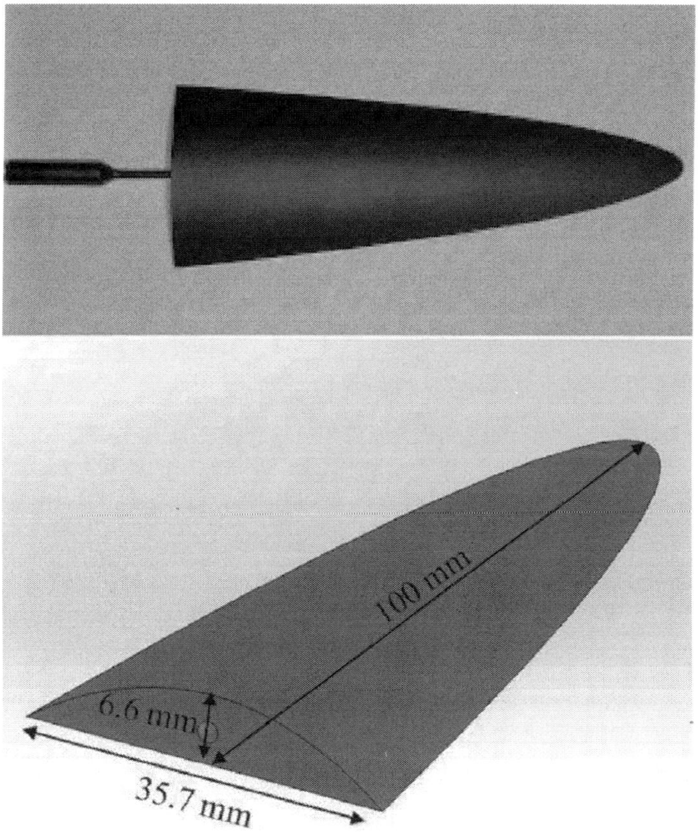

Figure 1.
3D CAO design of the waverider used in this experimental work.

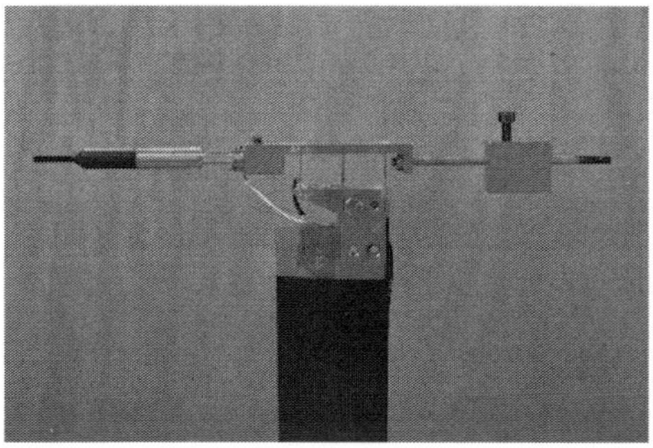

Figure 2.
View of the sting balance two modules, drag and lift.

lateral and vertical deformations. Indeed, other configurations such as a row of three wider slats showed lateral and vertical deformations that could weaken the balance of the sting, optimizing the bending which is the most important part for this module.

The drag slats are 0.2 mm thick and made of AISI 304 steel to avoid plastic deformation. Once all the components of the drag module are assembled, the lower

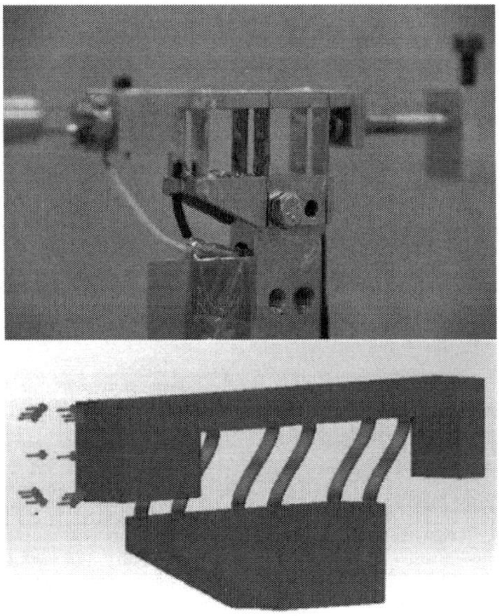

Figure 3.
Detail of the drag module and simulation of the strain stresses.

part is static and only the upper part is mobile. Therefore, the most sensitive part of the slats is the upper area, where the strain gauges will be placed. The strain gauges are glued to the central slat and there is one on each side to create a complete Wheatstone bridge. The lifting module is made up of a single part and there are only two superimposed slats. This module is also made of AISI 304 and it measures 10 cm. The purpose of having two superimposed slats is to reduce the bending of the heavy test models when the flow is deactivated. Indeed, if the test model creates a bending due to its mass, it can have an impact on the lift values in the presence of the flow. The maximum mass of the test model that this sting balance can support is 140 g. This lift module, like the drag module, has a full Wheatstone bridge with two strain gauges per side, as shown in **Figure 4**.

Drag module measurements are dependent on the position of the balance center of gravity (COG). A change in the position of the center of gravity will change the measurement of the associated forces for the same test model and flow conditions. To avoid this problem, a counterweight is placed at the back of the balance and its position is adapted according to the mass of the test model. The higher the mass of the test model, the further the counterweight will be from the drag module. On the contrary, if the test model is light, the counterweight will be closer so that the center of gravity is always aligned with the two slats equipped with strain gauges.

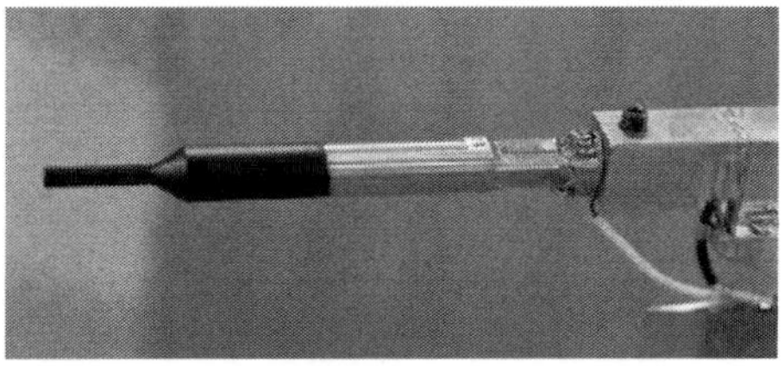

Figure 4.
Detail of strain gauges of the lift module.

Another important point for this type sting balance is to avoid that the flow has a direct impact on the deformation of the slats. Indeed, if the drag and lift modules are in the flow without protection, the drag and lift values will be overestimated. In addition to the shock produced by the presence of the test model, a second shock can be created on the drag module itself and increase the deformation of the slats and thus the drag value. To solve this problem, a cover was modeled and installed. It is dimensioned to allow the drag and lift modules to deform without touching the cover which could create friction and in turn distort the measurements. On the other hand, this cover also allows to protect the strain gauges from the temperature increase caused by the shock in hypersonic flow condition. For the experimental conditions of this work in supersonic regime, the shock does not produce any heating.

To validate this sting balance, we have performed a study with spheres in order to compare our experimental results with those of the literature. In his study, Aroesty determines the drag coefficients for spheres in a supersonic flow at Mach 4 and low density and establishes a curve that relates the drag coefficients of the spheres with the Reynolds number after shock Re2 [34]. In our case we measured the drag forces for spheres of different diameters in the three flow conditions at Mach 4. **Figure 5** plots the experimental data from Aroesti and our experimental results. As shown in the graph, the drag coefficients for the sting balance are consistent with the reference values. The numerous preliminary tests we have conducted have shown that without the stinger cover, the drag coefficients for small diameter spheres (less than 15 mm) and low Re2 (less than 200) are indeed overestimated as explained in the previous paragraph. For larger diameter spheres, the results are close to the Aroesty values without the protective cover. This is due to the fact that the shock created by the large diameter spheres is large enough to protect the scale elements that will be in the wake, which is not the case for the small diameter spheres.

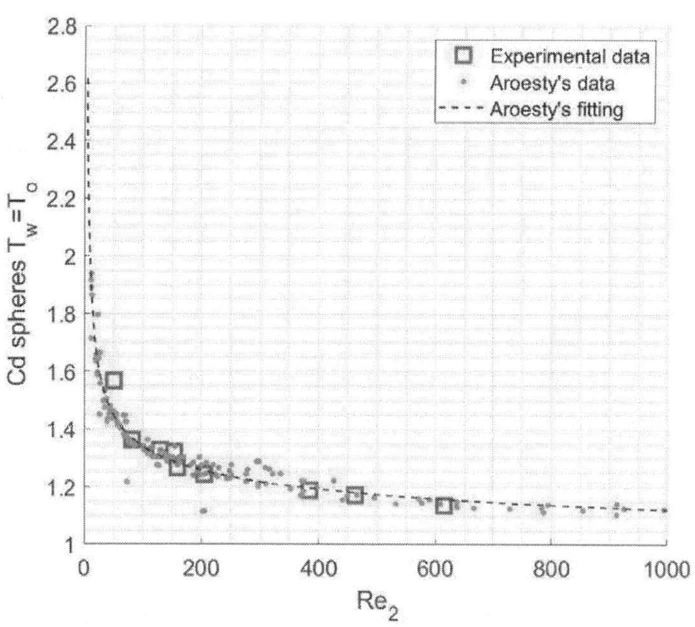

Figure 5.
Drag coefficients of spheres in near rarefied regime: Aroesty and Noubel data.

3.2 Strain gauges

Strain gauges are used to measure the slat deformation. It is a sensor whose resistance varies with the applied force, then it converts force into an electrical resistance which can then be measured [35]. Strain gauge elements are electrically connected to form a Wheatstone bridge circuit used for the measurement of static or dynamic electrical resistance. The output voltage of the Wheatstone bridge is expressed in millivolts output per volt input. At the end a calibration curve of the lift and drag modules will convert the out put signal in a mesure of the corresponding fonce expressed in newton. For the experimental conditions of the MARHy wind tunnel forces are estimated between 1 mN and 1 N so strain gauge need to have a high gauge factor. There are essentially two types of strain gauges, the so-called semiconductor strain gauges whose gauge factor is greater than 180 and the metallic strain gauges whose gauge factor is between 2 and 3. In order to optimize the measurement of forces we have opted for the first type, however they are very sensitive to the value of excitation applied which can easily damage them [36, 37]. For this balance, the gauge used is the KYUOWA: KSPB-2-1 K-E4, with an excitation voltage of 1 Volts.

3.3 Calibration procedure

Calibration of the balance is necessary to convert the electrical signal from by strain gauges into applied force. The purpose of calibration is to reproduce known forces on the sting balance and relate them to the variation of the measured voltage. We designed a calibration bench composed of a digital newton meter mounted on a motorized micrometric displacement, which aligned with the lift or drag modules, apply a stable and known force in the same geometric configuration as those of the experience. Both during calibration and during wind tunnel experiments, the models are screwed onto the model holding sting so as to align their center of gravity with the end of the sting. Finally a curve is obtained expressing the force as a function of the voltage measured for each module: drag and lift. The accuracy of the force measurement on the lift and drag module is estimated at 0.1 mN.

A counterweight is added so that the position of the center of gravity remains constant regardless of the size and mass of the model. This preserves the calibration performed with the balance provided that, during wind tunnel measurements, the balance's center of gravity is in the same position as during calibration. For this purpose, the position of the counterweight is adjusted for each test model.

It should be noted that if the aerodynamic forces are to be studied as a function of the angle of incidence of the test model with respect to the flow, it must be ensured that the center of gravity remains unchanged. For this reason, the balance is positioned horizontally so that the angle of incidence in the horizontal plane (x-y) is changed and the position of its center of gravity is not affected [38].

4. Experimental conditions

4.1 Marhy facility description

The first version of the wind tunnel, then named SR3, was built in 1961, at the Aerothermic Laboratory. Since 2000, this wind tunnel, renamed MARHy, has been installed at the ICARE Laboratory of the CNRS in Orléans, and many technical improvements have been made.

In particular the pumping unit which is a key element allowing to ensure in continuous mode without time limit low density flows with Mach numbers ranging from 0.8 to 21. The two main components of the Marhy facility are the large capacity test chamber and the pumping unit. The wind tunnel consists of three main parts: the settling chamber, the test chamber and the collector-diffuser chamber as presented in **Figure 6**.

The facility is supplied with exchangeable nozzles to generate laminar subsonic, supersonic and hypersonic stationary flows [39]. Only the supersonic configuration is described in this paper.

Subsonic and supersonic conditions flows are generated by contoured nozzles, using air or nitrogen at ambient temperature. Nozzles are housed into the settling chamber, a cylinder of 2.6 m length and 1.2 m in diameter with a large access port at the bottom (200 *kg*), placed on a trolley with wheels on a guide rail for easy opening and closing. The relevant components of the facility MARHy in supersonic configuration is skecthed on **Figure** 7.

The divergent section of the nozzles opens into the experimental chamber, a cylindrical test section with a diameter of 2 m and a length of 3.5 m.

This chamber is placed perpendicular to the direction of the flow, so that there is enough space on either side of the flow to install diagnostic systems and supports without disturbing the flow. The experimental chamber is large enough to allow the

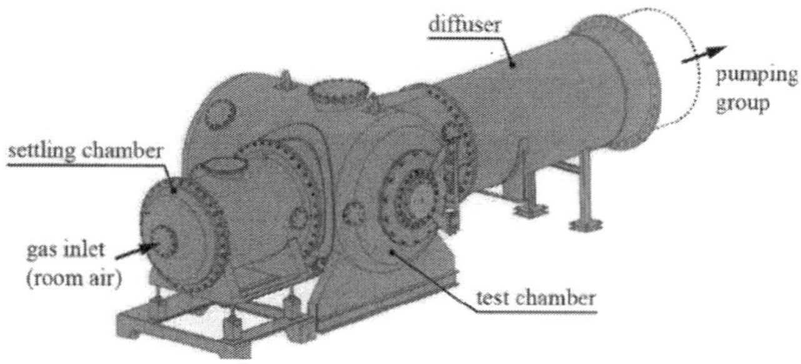

Figure 6.
The wind tunnel MARHy and the pumping group.

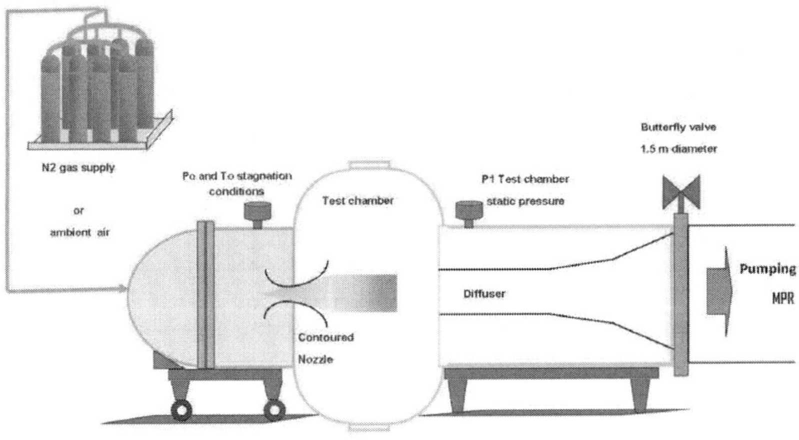

Figure 7.
Sketch of the wind tunnel MARHy in supersonic configuration.

integration of specific instrumentation such as probe supports, electron gun and aerodynamic balances and to avoid interactions between the flow and the wall of the wind tunnel. Two rounded bottoms laterally close the cylinder, one of them is provided with a door of 1.2 m in diameter, giving access to the interior of the experimental chamber as can be seen in **Figure 8**. Four port flanges with a diameter of 0.6 m, closed by optical windows made of quartz and fluorine, are distributed around its cylindrical section. Six other smaller diameter flanges are available for instrumentation.

The pressure in the experimental chamber is recovered by means of a central diffuser to match the test section pressure to the inlet pressure for the vacuum pumps, specially at hypersonic operating conditions. The collector-diffuser is a 1.4 m diameter cylinder connects the experimental chamber to the pumping group by means of a motorized butterfly valve with a diameter of 1.5 m. The pumping group consists of 2 primary pumps, 2 roots-type intermediate pumps and 12 roots-vacuum pumps,ensuring continuous operation. The number of pumps commissioned depends on the operating conditions of the flow (Mach number and static pressure).

Figure 8.
View of the experimental chamber of the facility MARHy.

4.2 Operating conditions

One of the major advantage of the facility is its almost unlimited running time, particularly suitable to obtain a good stabilization of the flow conditions prior to experiments.

The purpose of this experimental study is to investigate the influence of rarefaction effects on the flight performance of waveriders with isobar and isomach operating conditions. Among the available nozzles we have selected four to carry out this first part of the study in supersonic conditions. The Isobar conditions are obtained with a static pressure of 8 Pa with the nozzle N1 operating at Mach 2 and nozzle N3 operating at Mach 4. Isomach conditions are performed at Mach 4 with the nozzle N2 operating with a static pressure of 2.66 Pa, the nozzle N3 with a static pressure of 8 Pa and the nozzle N4 operating with a static pressure of 71 Pa. The detailed operating conditions for these nozzles are presented in **Tables 1–4** where p represents the pressure, T the temperature, ρ the volumetric mass density, U the flow velocity, M the Mach number, λ the mean free path and R_e the Reynolds number based on the length L = 100 mm and calculated with the relation $R_e = U_1 L/\mu_1$, where μ_1 is the kinematic viscosity.

Various parameters have been proposed to quantify the rarefaction effects that make the Navier–Stokes equations invalid. The Knudsen number, defined as the ratio of the mean free path to a characteristic length, is a measure of the gas density

Stagnation conditions	Free stream conditions
$p_0 = 63$ Pa	$p_1 = 8$ Pa
$T_0 = 293$ K	$T_1 = 163$ K
ρ_0=7.44.10^{-4} kg.m^{-3}	ρ_1=1.71.10^{-4} kg.m^{-3}
	μ_1=1.10^{-5} Pa.s
	$U_1 = 511$ m.s^{-1}
	$M_1 = 2$
	$\lambda_1 = 0.278$ mm
	$R_e = 77.3$ /cm

Table 1.
N1 (Mach 2–8 Pa) operating conditions.

Stagnation conditions	Free stream conditions
$p_0 = 405$ Pa	$p_1 = 2.66$ Pa
$T_0 = 293$ K	$T_1 = 70$ K
ρ_0=4.81.10^{-3} kg.m^{-3}	ρ_1=1.33.10^{-4} kg.m^{-3}
	μ_1=4.77.10^{-6} Pa.s
	$U_1 = 670$ m.s^{-1}
	$M_1 = 4$
	$\lambda_1 = 0.219$ mm
	$R_e = 180.5$ /cm

Table 2.
N2 (Mach 4–2.66 Pa) operating conditions.

Stagnation conditions	Free stream conditions
$p_0 = 1214$ Pa	$p_1 = 8$ Pa
$T_0 = 293$ K	$T_1 = 70$ K
ρ_0=1.44.10^{-2} kg.m^{-3}	ρ_1=3.99.10^{-4} kg.m^{-3}
	μ_1=1.10^{-5} Pa.s
	$U_1 = 670$ m.s^{-1}
	$M_1 = 4$
	$\lambda_1 = 0.072$ mm
	$R_e = 541.5$ /cm

Table 3.
N3 (Mach 4–8 Pa) operating conditions.

Stagnation conditions	Free stream conditions
$p_0 = 10797$ Pa	$p_1 = 71$ Pa
$T_0 = 293$ K	$T_1 = 70$ K
ρ_0=0.13kg.m^{-3}	ρ_1=3.55.10^{-3} kg.m^{-3}
	μ_1=4.77.10^{-6} Pa.s
	$U_1 = 670$ m.s^{-1}
	$M_1 = 4$
	$\lambda_1 = 0.008$ mm
	$R_e = 4806.7$ /cm

Table 4.
N4 (Mach 4–71 Pa) operating conditions.

and is in fact a state parameter. Other correlation parameters must be taken into account for rarefied high velocity flows as indicators of when the Knudsen number is not a flow parameter for supersonic flows in the sense that Mach number and Reynolds number are flow parameters. To account for viscous effects due to rarefaction and high velocity flow, the similarity number $\psi = \frac{M}{\sqrt{Re}}$, called the viscous parameter is a better candidate where the Reynolds number reflects the dynamical similarity of the flows around a model and the full-scale object [40–42]. It is important to note that in this investigation the wall temperature of the waverider remains "cold", with $T_{wall} = T_0$.

Table 5 sumarizes the values for the Knudsen number, the Reynods number, and the viscous parameter for each of the nozzles used for this investigation.

Parameter	Mach2-8 Pa	Mach4-2 Pa	Mach4-8 Pa	Mach4–71 Pa
Kn	2.78.10^{-3}	2.09.10^{-3}	7.33.10^{-4}	8.25.10^{-5}
Re	773.8	1805	5415	48067
Viscous parameter	0.227	0.0943	0.0544	0.0182
Re2	514.84	513.15	1539.46	13687.61

Table 5.
Nondimensional parameters for each one of the nozzles (based on L = 10 cm).

4.3 Measurements protocole

The supersonic nozzles of the wind tunnel are curved nozzles which will produce isentropic flows by adjusting the generating pressure P_0 and the static pressure P_1 measured in the test section. The latter will be adjusted by adjusting the opening of the motorized butterfly valve which separates the test chamber from the pumping unit.

The test model is positioned in the isentropic zone, where the flow conditions are stabilized as showed on **Figure 9**. Outside this zone, the Mach number and pressure change and no longer correspond to the isentropic values for each nozzle. The total force measured by the balance corresponds to the force exerted by the flow as well as that created by the suction of the pumping unit. In order to keep only the force created by the flow, a measurement of the residual pumping force is made by isolating the model/balance assembly from the flow. To do this, a plate mounted on a rotating piston is inserted into the flow in front of the model, so that the model is no longer subject to the aerodynamic forces generated by the flow. The force of interest is the difference between the total force and the residual force.

The aerodynamic force measurements are done in two steps: first, the values of the drag and lift modules corresponding to the waverider exposed in the flow are recorded, then in a second step, the plate is placed between the waverider and the flow to record the offsets of both modules. For both steps, the acquisition time is 10 seconds with an acquisition frequency of 1000 kHz. This operation is repeated 5 times for each position of the wave rider in the flow, in order to have an average value which is representative of the aerodynamic forces. Since the isentropic flow is homogeneous in a given volume which depends on the nozzle used, special attention has been paid to the placement of the waverider in the flow whatever the angle of incidence. For this purpose, the position of the waverider is corrected using motorized translations, so as to reposition the waverider to its initial position without angle of incidence as illustrated on **Figure 10**. The reference point is the center

Figure 9.
Assembly of the sting balance/model in the test chamber.

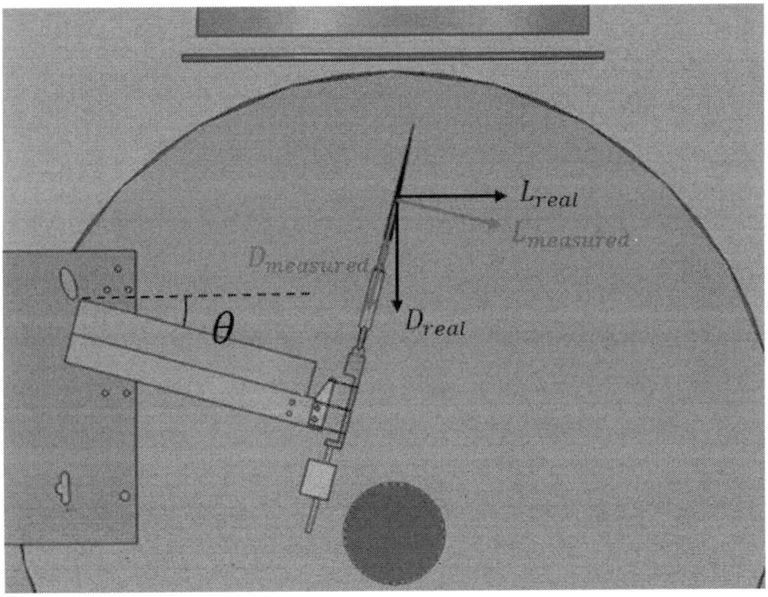

Figure 10.
Coordinates correction of the model location in (x-y) plane.

of the waverider. The determination of the drag and lift forces is a function of the measured values of the respective modules and the angle of attack of the waverider. The correction is given by changing the Cartesian coordinates into cylindrical coordinates following these equations:

$$Drag_{real} = Drag_{measured} * cos\,(\Theta) + Lift_{measured} * sin\,(\Theta) \quad (1)$$

$$Lift_{real} = Lift_{measured} * cos\,(\Theta) - Drag_{measured} * sin\,(\Theta) \quad (2)$$

5. Results

5.1 Measurements of lift and drag forces

The lift-to-drag ratio is of most important consideration when designing spacecraft and others space vehicles to reach higher lift-to-drag ratios. In addition to their shape, the incidence angle plays a significant role over the lift-to-drag ratio. The incidence angle is defined as the angle at which the leading edge of the vehicle is set in relation to the flowing air, and has a direct influence on how far the vehicle will glide for a given altitude. At high Mach numbers or high altitudes because of viscous effects, the skin friction will increase the viscous drag, decreasing L/D ratios.

Drag and lift forces have been measured for each one of the nozzles and for several angle of attack covering negative and positive angles. The objective is to study the behavior of the waverider in gliding phase, maximizing the Lift-to-Drag ratio (positive angles) and in the landing phase, reaching a target as quickly as possible (negative angles). For nozzles N1 and N2, the attack angles range between −25° and 25° while for N3 and N4 angles are ranged between −50° and 25°. It was not possible to achieve angles below −25° with nozzles N1 and N2 because the vacuum pumps are not powerful enough to achieve the required pressure P_1. **Figures 11** and **12** plots the lift and drag forces measured for each operating

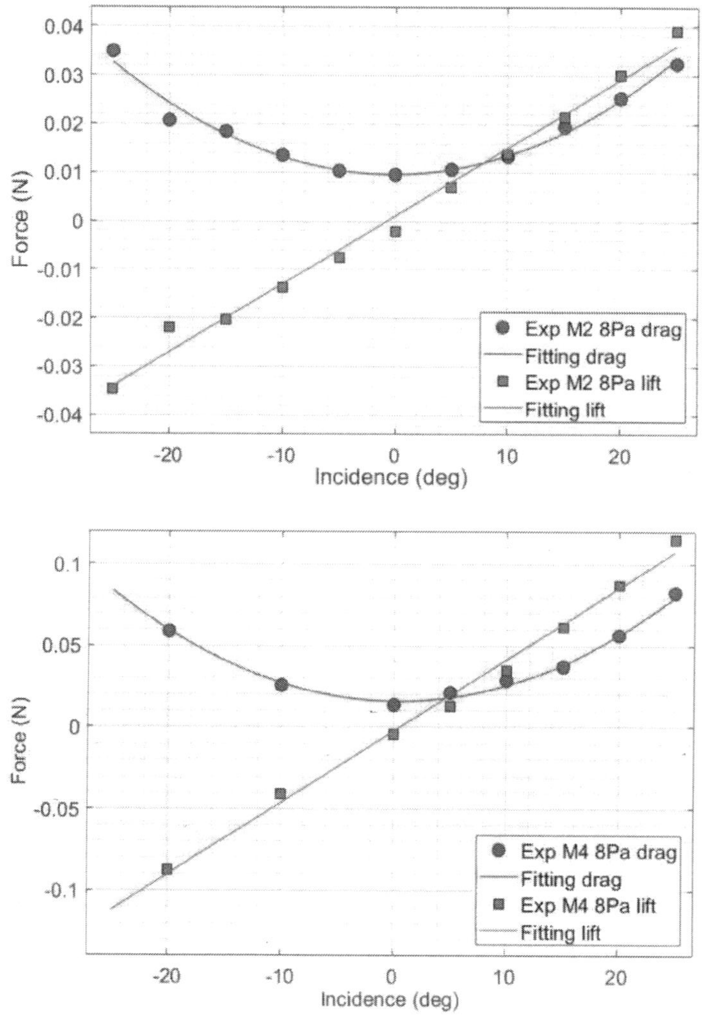

Figure 11.
Lift and drag forces: Mach 2–8 Pa and Mach 4–8 Pa.

condition. Whatever the experimental condition, the lift force behaves linearly with the angle of attack Θ, while the drag force follows a quadratic behavior at order 2. Lift forces can be described with an equation at first order as:

$$Lift = a\Theta - b \tag{3}$$

The fitting coefficients a and b are sumarized on **Table 6**.

Concerning the drag forces, an equation in second order can describe their evolution with the angle of attack, as follows:

$$Drag = c(\Theta)^2 + d\Theta + e \tag{4}$$

Table 7 presents the corresponding coefficients c,d and e.

The y-intercept is negative due to the shape of the waverider, the asymmetry between the two surfaces creates a negative lift at 0° angle of attack. Negative lift values means that the force acts in the same direction as the gravity force.

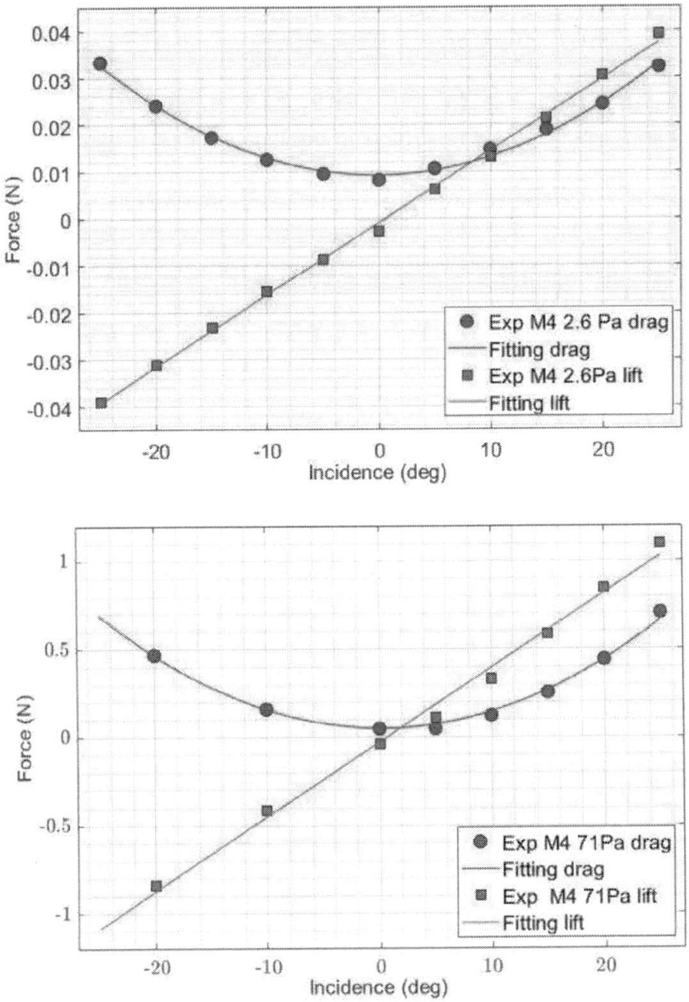

Figure 12.
Lift and drag forces: Mach 4–2 Pa and Mach 4–71 Pa.

Nozzle	a	b
Mach 2–8 Pa	$1.4.10^{-3}$	1.10^{-3}
Mach 4–2 Pa	$1.53.10^{-4}$	$8.63.10^{-4}$
Mach 4–8 Pa	$4.39.10^{-3}$	2.52^{-3}
Mach 4–71 Pa	$4.22.10^{-2}$	2.74^{-2}

Table 6.
Fitting coefficients for lift forces.

Comparisons of lift and drag forces show that those measured at the operating condition Mach 4–71 Pa are the highest and on the contrary, the lowest forces are obtained for Mach 2–8 Pa and Mach 4–2 Pa which are close in terms of values and shapes. One of the major parameters to design a space craft is the lift-to-drag ratio value. This value may determine the gliding ability of the waverider. The L/D ratio is a dimensionless number that can be determined directly from the aerodynamic

Nozzle	c	d	e
Mach 2–8 Pa	$3.74.10^{-5}$	$1.6.10^{-5}$	$9.56.10^{-3}$
Mach 4–2 Pa	$3.78.10^{-5}$	$1.02.10^{-5}$	$9.24.10^{-3}$
Mach 4–8 Pa	$1.05.10^{-5}$	7.53^{-5}	$1.57.10^{-3}$
Mach 4–71 Pa	1.10^{-5}	5.76^{-5}	$5.38.10^{-3}$

Table 7.
Fitting coefficients for drag forces.

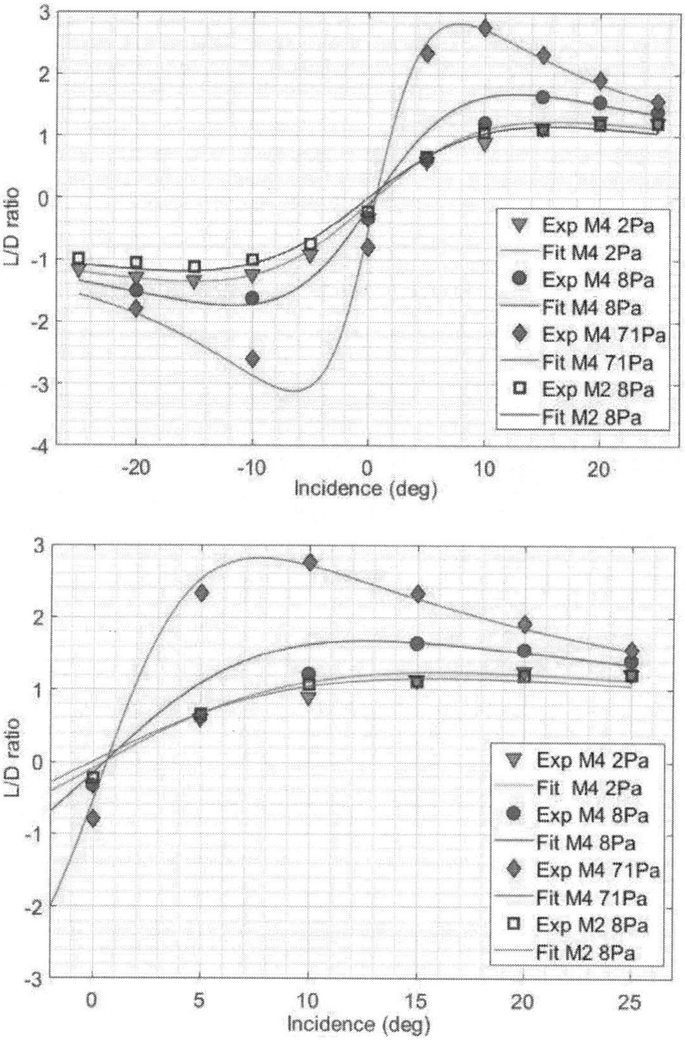

Figure 13.
Comparison of L/D ratio for all experimental conditions.

forces but also from the lift and drag coefficients. It does not follow a linear or quadratic function but has a more complex form. **Figure 13** presents the evolution of the ratio lift-to-drag with the angle of attack Θ for the overall experimental conditions. As can be seen, the L/D ratio will increase to reach a maximum and then decrease and converge to 0.5–0.6. This evolution occurs for both positive and negative angles and presents a quasi symmetry around the 0° angle. This symmetry is possible because the thickness of the waverider is small, 6.6 mm, compared to its length, 100 mm, so that the waverider can be compared to a flat beveled plate. The waverider can be rotated 180° lengthwise and the sliding performance will be similar or even better if the fitting curves are realistic. Negative angle behavior shows that the optimal landing angle is between −5° and − 15°, depending on the ambient operating pressure of the nozzle used.

In the following, only the results of positive angles that illustrate the ability of the waverider to glide and maximize the travel distance before landing will be analyzed. The analysis of the results shows that the angle for which the value of the L/D ratio is maximum changes as a function of the operating conditions of the nozzle i.e. as a function of the static pressure P_1 and the Mach number. By relating the pressure P_1 of the flow to the atmospheric pressure for a given flight altitude, it is possible to associate to each nozzle an equivalent flight altitude, 80 km for Mach 4–2 Pa and Mach 2–8 Pa, 70 km for Mach 4–8 Pa and 50 km for Mach 4–71 Pa. This means that the angle of the maximum value of the L/D ratio decreases with altitude while the value itself increases. This is due to the fact that when the pressure decreases and the speed remains constant, the waverider must adapt its incidence in order to maintain an optimal L/D ratio and travel as far as possible. The influence of pressure in the isomach condition is shown in **Figure 14**. The values of the L/D ratio vary from 2.8 and 1.2 respectively for angles varying from 5° and 15°.

This observation shows that the waverider needs a high angle of attack at the beginning of its flight at high altitude, (low pressure) and must decrease its incidence to have an optimal L/D ratio when it decreases its flight attitude. A non-optimized L/D ratio will decrease the range of its flight. For example, if the waverider is launched at 80 km altitude with an incidence of 10° the L/D ratio is not optimal and for a flight altitude between 80 km and 50 km the waverider will be able to travel 33 km. Indeed between 80 km and 70 km, by referring to the curve for the Mach 2-8 Pa nozzle the value of L/D is of 0.89 and between 70 km and 50 km the M4 8Pa curve gives a L/D value of 1.21. However, by varying the angle of incidence the optimal L/D ratio between 70 km and 80 km is 1.25 and 1.64 for an

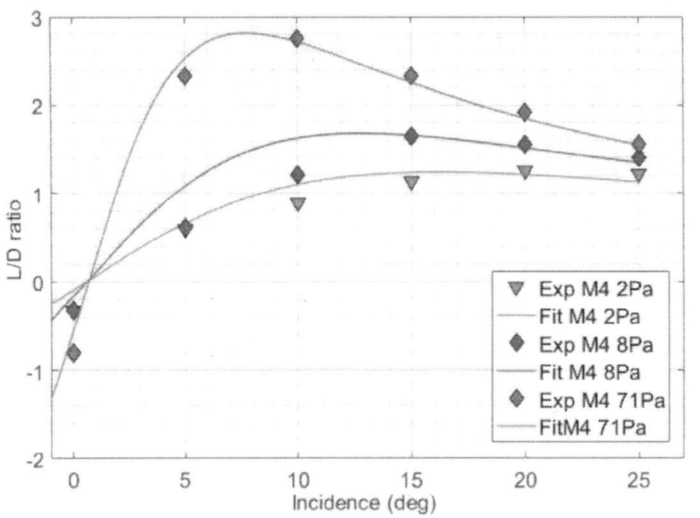

Figure 14.
Pressure influence on L/D ratio for isomach experimental conditions(Mach 4).

altitude ranged between 70 km and 50 km. With these data the new maximum flight distance that the waverider will be able to cover is 43 km. There is a variation of 10 km, which is not much, but during the last 50 km the L/D values could increase significantly due to the increase of the atmospheric pressure. The variation of the distance flown between the optimized and non-optimized flight can be greater, in the order of 50 km to 100 km. During the re-entry into the atmosphere, the speed of the waverider will vary and increase while decreasing the altitude. In this work, Mach 2 and Mach 4 nozzles operating with a static pressure of 8 Pa were used to observe the influence of the speed on the aerodynamic performance of the waverider. The L/D ratio is presented in **Figure 15**. As observed, the L/D ratio increases with Mach number as well as its optimal angle of attack to a lesser extent. For the Ma 2 8 Pa curve, the maximum is reached at about 25° while for the Ma 4 8 Pa curve the optimal angle is 15°. For the same pressure, when the speed increases, the value of the L/D ratio also increases. The maximum value of the L/D ratio can reach higher values for smaller angles, so to optimize the L/D ratio of waveriders, their speed must be increased when they fly at high altitude.

A first conclusion can be drawn from these experimental results. To optimize the performance of the waverider, the L/D ratio must be maintained at its optimal value by increasing the Mach number at high altitude and by increasing the angle of incidence by decreasing the flight altitude, in order to glide over the greatest possible distance. For the test model of this work the incidence will be between 20° and 25° at 80 km and will gradually decrease to 8° at 50 km. In terms of performance, the waverider will travel 1.2 km for every vertical kilometer lost at 80 km altitude and at 50 km, the waverider will travel more than 3 km. In fact, a waverider geometry is optimized for a specific flying speed and pressure. The geometry of this waverider is optimized to fligh at Mach 10 and altitudes ranging between 40 km and 50 km as Rolim presented in his paper, giving higest L/D values.

5.2 DSMC simulations and comparison with experimental results

Numerical simulations have been carried out using the direct Monte Carlo method, a stochastic method, solving the flow fields in transitional regime i.e. from continuum low density regime to free molecule regime. For this purpose, the

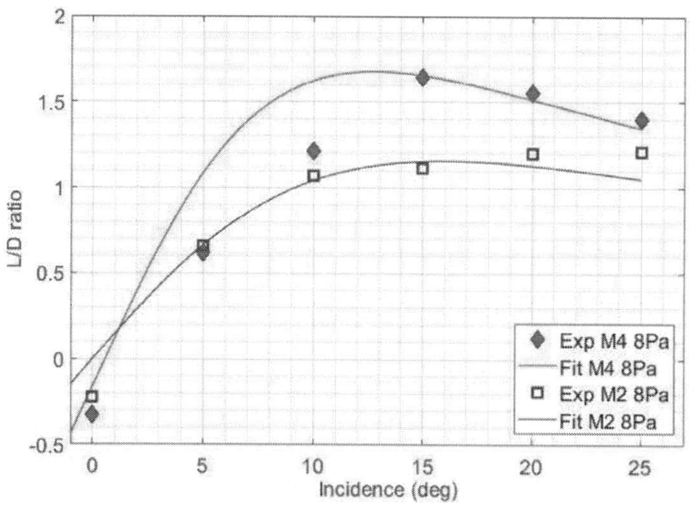

Figure 15.
Mach number influence on L/D ratio for isobar experimental conditions (P_1=8 Pa).

well-known DS3V software developed by Prof. Graeme A. Bird have been used [43, 44]. The main objective is to determine the accommodation parameters between the gaz flow and the waverider surface for our experimental conditions and analyze how they are related to rarefaction effects.

The comparison criterion between the numerical and experimental results are the drag and lift forces with the purpose to obtain numerical results as close as possible to those of the experiments.

We have selected the DS3V program by many reasons. The first one is the avaibility as it is free to download from the Bird website (http://www.gab.com.au). This program was designed to be able to run on personal computers with Microsoft windows. Finally, the use of DS3V is quite simple for a beginner as it uses a set of menus which creates files containing all the simulation information for post processing. However, one disadvantages associated with the DS3V codes is the creation/importation of the 3D geometry. Indeed, DS3V does not have an integrated geometry package and does not accept basic 3D CAD files. The geometry file have to be a Raw Triangle files with a series of x, y and z coordinates which form a 'triangle' and when put together they form a mesh.

To design the waverider, we have used the software Rhinoceros 7, because it can create triangular mesh with coordinates associated to each triangle apex. It is possible to select gas-surface interaction models among two models: the diffuse model or the CLL model [45–47]. The diffuse reflection model was adopted for this study. However, the reflection and absorption parameters characterizing the interaction with the waverider surface were adapted to the experimental conditions to be simulated. As the DSMC simulation is a probabilistic calculation, we considered that a flow time of 1 ms was sufficient to obtain valid results, and in particular this time is sufficient for the aerodynamic forces to be stabilized. Simulation were realized allowing 200 Mb memory. **Figure 16** shows the result obtained with the DS3V code

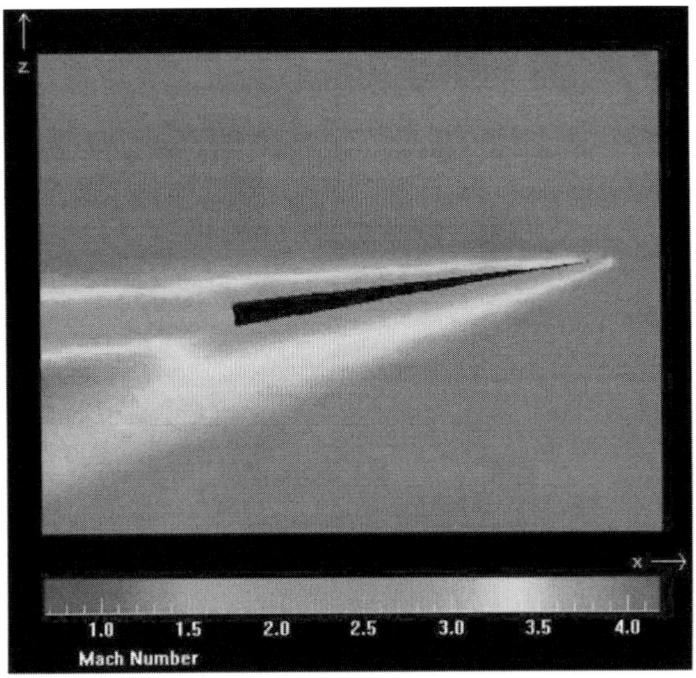

Figure 16.
Illustration of a numerical result obtained with DS3V for the experimental conditions Mach 4–8 Pa.

simulating the waverider with an incidence of 10° with conditions similar to those obtained experimentally with a Mach 4–8 Pa flow.

To achieve similarity between numerical and experimental curves, the accomodation coefficients which correspond to the specular reflection fraction and the absorbed fraction at surface are adapted. The sum of both fraction need to be equal to 1 for a consistent simulation. These coefficients are summarized in **Table 8** for each nozzle.

Parameter	Mach 2–8 Pa	Mach 4–2.66 Pa	Mach 4–8 Pa	Mach 4–71 Pa
Reflection	0.825	0.86	0.90	0.95
Absorption	0.175	0.14	0.10	0.05

Table 8.
Reflection and absorption accomodation parameters.

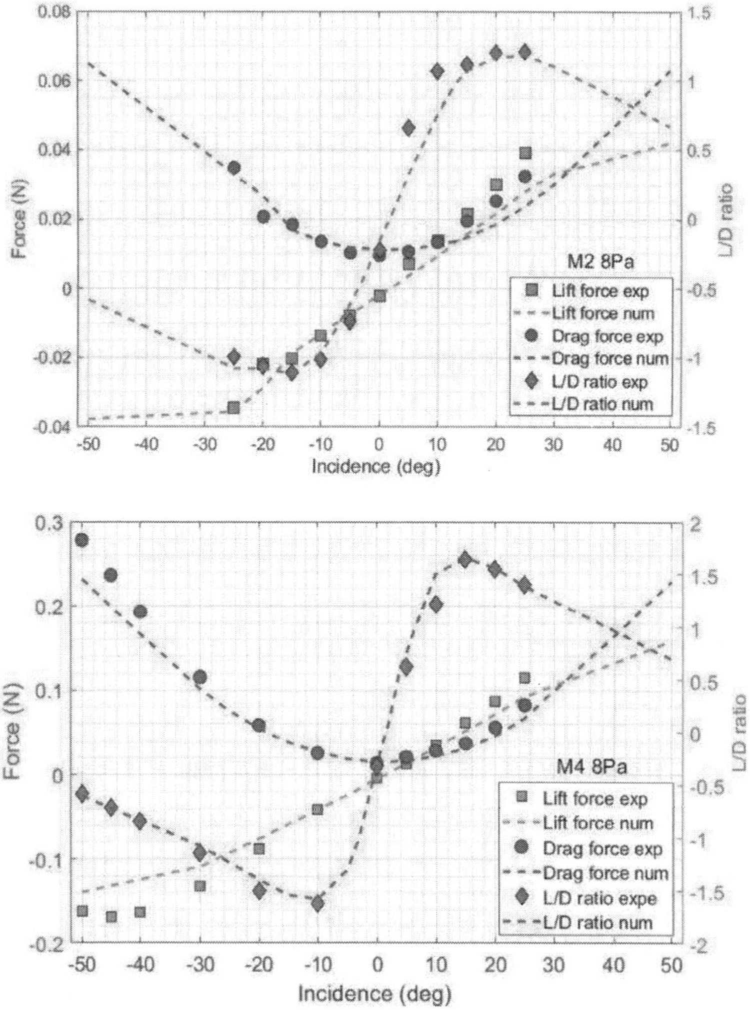

Figure 17.
Comparison of experimental and DS3V results for Mach 2–8 Pa and Mach 4–8 Pa.

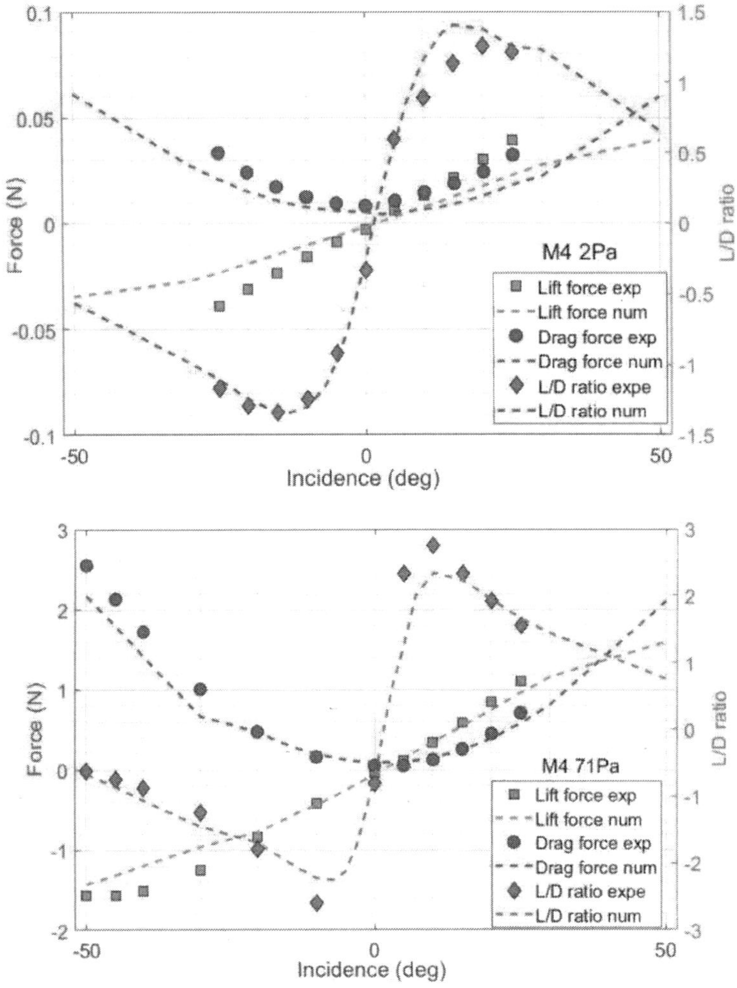

Figure 18.
Comparison of experimental and DS3V results for Mach 4–2.6 Pa and Mach 4–71 Pa.

For each of the experimental conditions i. e. nozles, the simulation of the flow around the wavereider was performed for the same incidences as those tested experimentally. Comparison between experimental and numerical results are presented on **Figures 17** and **18**. One can observe à good correlation between the numerical and the experimental results. Nevertheless, there are some differences in particular for the case Mach 2–8 Pa and Mach 4–2 Pa wich are the more rarefied conditions and presents small force values. The coefficient of the specular reflexion parameter seems to decrease with the rarefied parameter (Kn, $\psi = \frac{M}{\sqrt{Re}}$ or Re2). For low altitudes, typically the condition Mach 4–71 Pa, the reflection parameter tends to 1.

6. Discussion

6.1 Influence of rarefaction effects on fligth performance

The optimal flight conditions are given by the maximum values of the drag-lift ratio. These values will depend on the effects of rarefaction as summarized in

Table 9 in which are reported the maximum values of the ratio drag-lift, the angle of incidence corresponding for the various conditions of experiment studied in this work. Similarly, for each case, the values of the accommodation coefficient optimized in the framework of the DSMC numerical simulations are reported.

Figure 19 shows the maximum value of the L/D ratio and the corresponding angle as a function of the Knudsen number. It can be seen that the variation of the optimum angle of incidence increases linearly with the number of Knudsen, while the value of the L/D ratio decreases linearly with the logarithm of the number of Knudsen and this without distinction of the Mach number.

The study of the evolution of the ratio L/D with the rarefaction parameter, shows the influence of the Mach number as illustrated in **Figure 20**. Indeed, both the maximum value of the ratio L/D and the corresponding angle of incidence follow a linear variation for the experimental conditions at Mach 4, while the values for Mach 2 do not follow this trend. Isobar results shows that lower Mach number needs higher angle of incidence to optimize de ratio L/D.

As shown in **Figure 21**, the reflection accommodation coefficient used for the DSMC numerical simulations also follows a linear trend with the Knudsen number and increases towards 1 with decreasing Knudsen number. The same conclusions are made with the DSMC parameter (reflection parameter). However, when the comparison is made with the rarefaction parameter, only the results obtained with the same number of mach follow a linear trend.

In conclusion, for each parameter, the Knudsen number regroup all nozzles on a same trend (linear or logarithmic. It is not the case for the rarefaction parameter but a study isomach seems possible. To valid these conclusions, studies in hypersonic Mach 20 will be realized.

6.2 Estimation of the skin friction contribution

'The different effects of Mach number and geometry optimization on the aerodynamic performance of waveriders have been demonstrated. However, the effects of viscous drag, which have been little studied, can also have an impact on an elementary waverider geometry. For comparison purposes, and to complete our results, we consider the results presented by Rasmussen, carried out with a Mach number of 4 and with a geometry similar to the one used for the present study [48].

The ratio L/D can be expressed as:

$$L/D = \frac{L}{D_0 + D_f + D_b} \tag{5}$$

where L is the lift force, D_0 is the invicid wave drag, D_f is the friction drag and D_b is the base drag. In the following we assume that the base pressure is equal to the freestream pressure so that the base drag can be negleted [29]. The friction drag can be written as follows:

$$D_f = q S_w C_f \tag{6}$$

Parameter	Mach 2–8 Pa	Mach 4–2.66 Pa	Mach 4–8 Pa	Mach 4–71 Pa
angle of maximum L/D ratio	25	20	12.6	7.8
L/D ratio maximum	1.2086	1.2538	1.677	2.816
Reflection parameter	0.825	0.86	0.90	0.95

Table 9.
Reflection and absorption parameters.

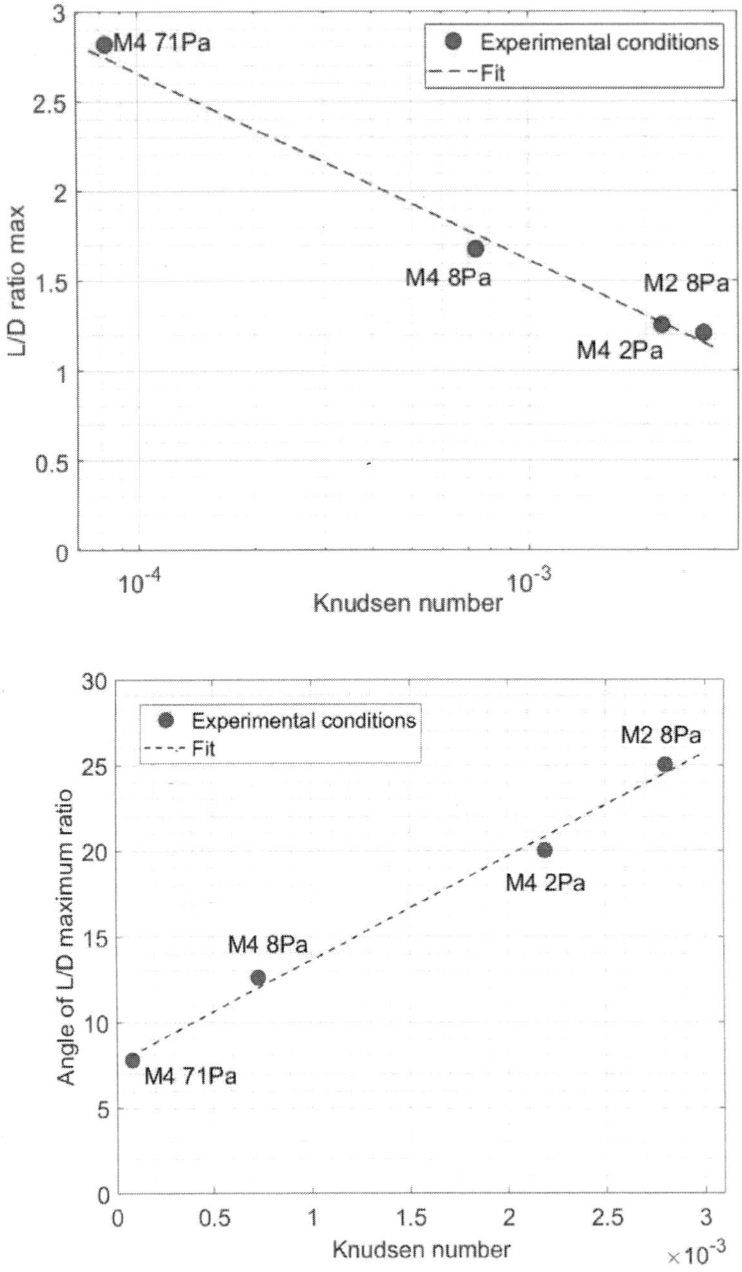

Figure 19.
Value of maximum L/D ratio and value of the corresponding angle of incidence as a function of the Knudsen number.

where q is the dynamic pressure, S_w is the wetted area of the waverider and C_f is the friction coefficient.

Friction effets are related to the development of the boundary layer, that can be influenced by many factors. The friction coefficient may depend of such factors as the Mach and Reynolds numbers, the wall temperature, and in continuum regime others physical properties such turbulence and flow

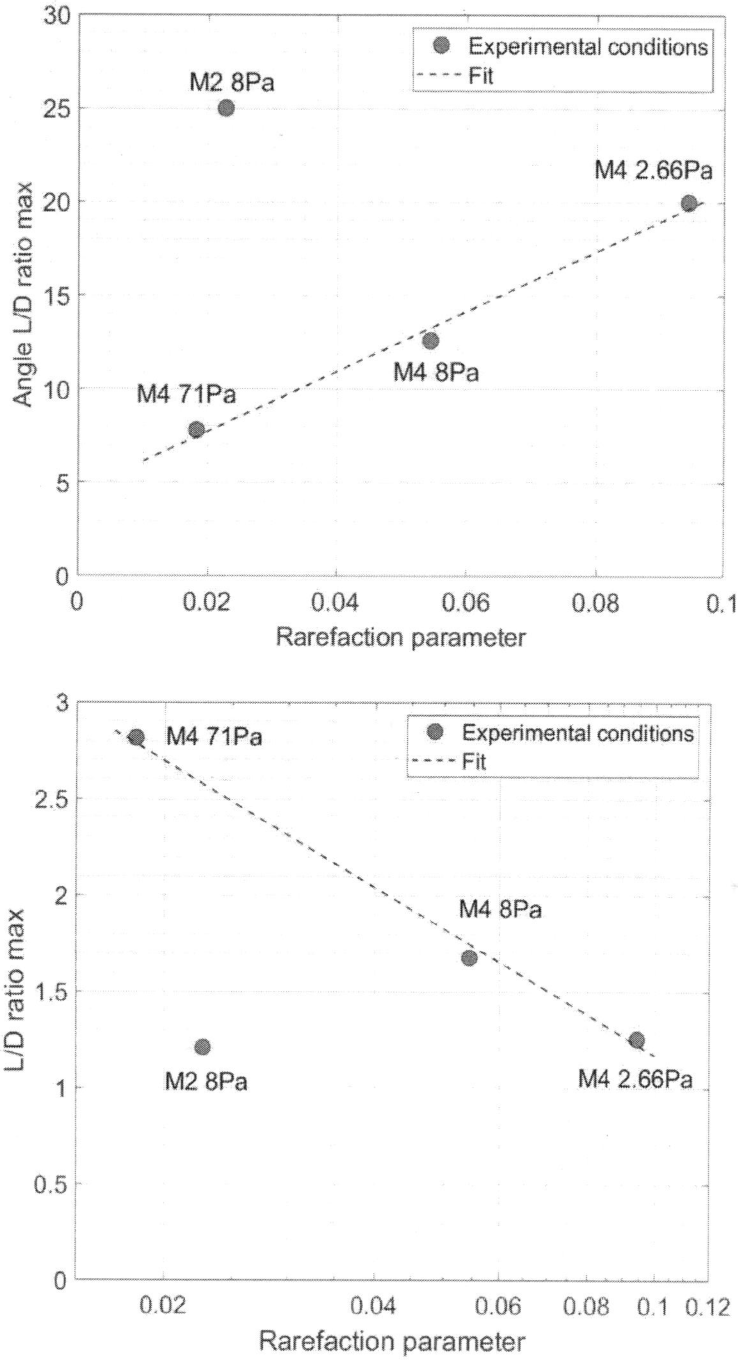

Figure 20.
Value of maximum L/D ratio and value of the corresponding angle of incidence as a function of rarefaction parameter value.

separation. This means that the definition of the friction coefficient can be a complex function that strongly depends on the flow properties and the objet model geometrie. For the current study, the waverider geometrie can be

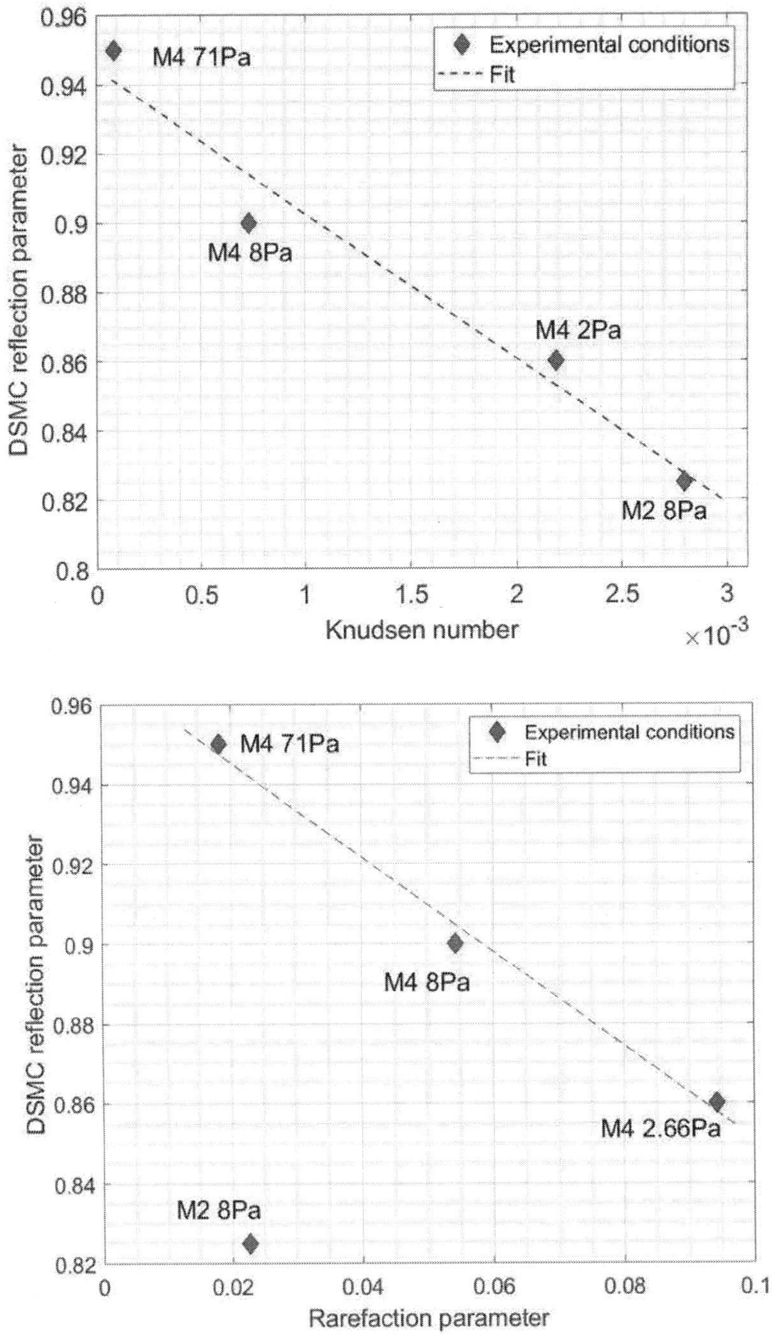

Figure 21.
Evolution of the DS3V reflection parameter as a function of the Knudsen number and the rarefaction parameter.

assimilated to a sharp plate, without the formation of a bow shock at the leading edge. Under this assumption, the skin friction can be defined by the following equation:

$$C_f = \frac{1.328 f(M, T_w/T_\infty, Re)}{\sqrt{\frac{\rho_\infty V_\infty l}{\mu_\infty}}} \tag{7}$$

Where $f(M, T_w/T_\infty, Re)$ is a function depending on Mach number, wall temperature and Reynolds number. For simplicity this function value can be assumed to be equal to 1 for $T_w/T_\infty = 1$, wich is our case. For the present study, the subscribe inf corresponds to the values of the free stream noted with the subscribe 1 in **Tables 1–4** presenting our experimental conditions.

Table 10 summarizes values of the friction coefficient for the experimental conditions presented by Rasmussen and those of the present experimental work as well as the maximum value of L/D ratio and the corresponding incidence angle of

Condition	C_f	$V^{2/3}/S\cos\alpha$	Max (L/D ratio)
Mach 2–8 Pa	0.047	0.2074	1.2
Mach 4–2 Pa	0.030	0.2	1.25
Mach 4–8 Pa	0.0178	0.196	1.67
Mach 4–71 Pa	0.006	0.189	2.8
Rasmussen	0.001	0.12	8
Rasmussen	0.002	0.15	5.9
Rasmussen	0.003	0.175	4.2

Table 10.
Sumarize of the parameters for the optimum L/D ratio corresponding to the experimental conditions of this work and those predicted by Rasmussen.

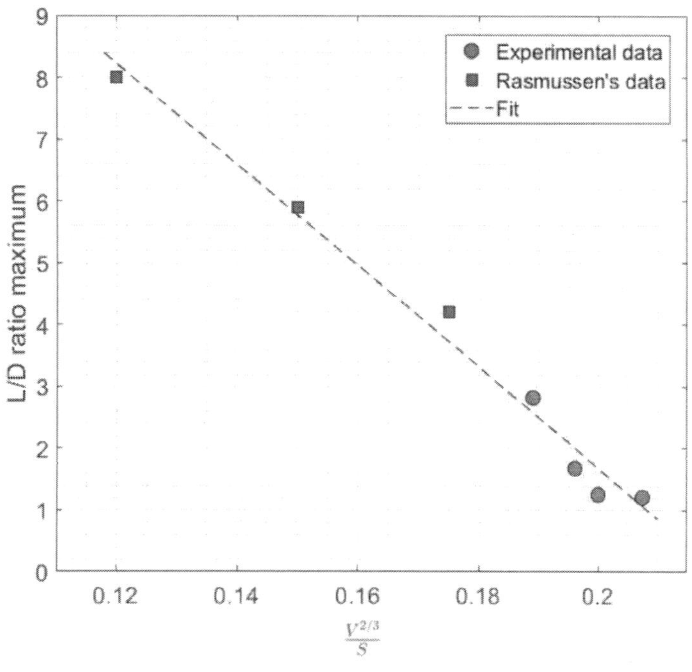

Figure 22.
Evolution of the maximum value of L/D as a function of the volume ratio.

the waverider. The value $V^{2/3}/(S\ cos\,\Theta)$, corresponds to the volume ratio wich is a function of the incidence angle Θ.

Figure 22 plots the Rasmussen data for the Mach 4 case and our experimental resuls which presents the linear evolution of the maximum L/D value with the

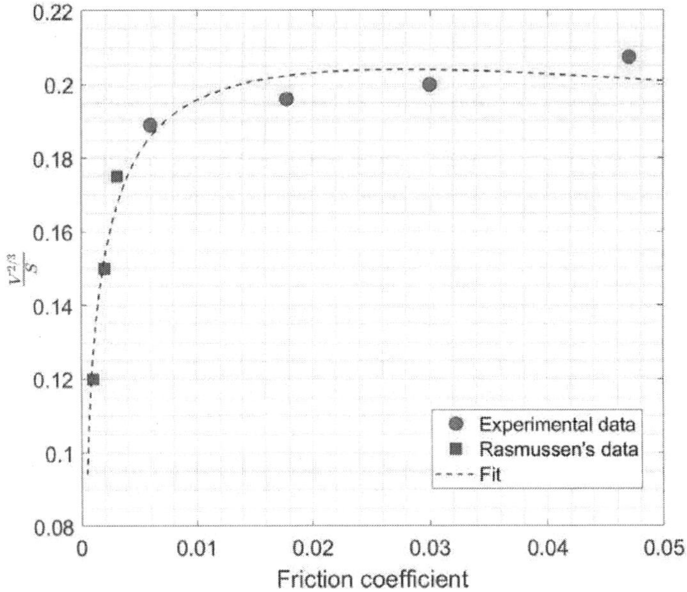

Figure 23.
Evolution of the volume angle as a function of the friction coefficient.

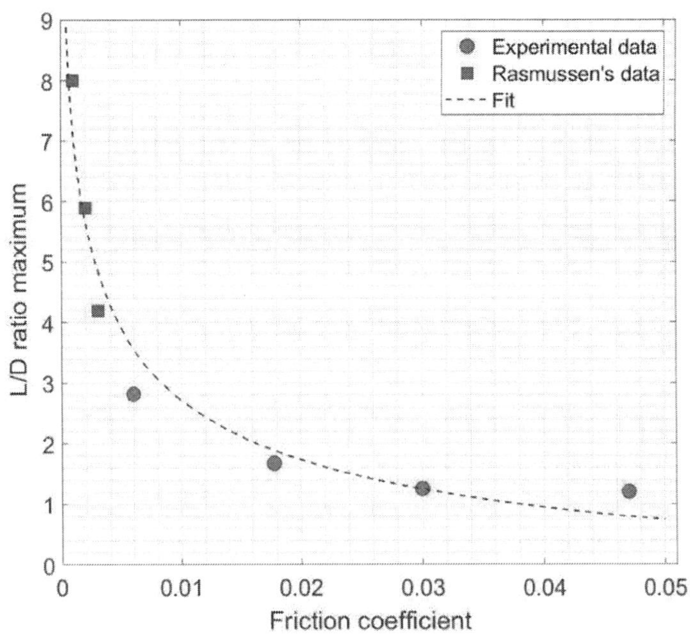

Figure 24.
Evolution of the maximum value of L/D as a function of the friction coefficient.

volume ratio. These results confirm the increase of the L/D maximum value as approaching continuous regime, the Rasmussen conditions presents lower friction coefficient values compared to those of our experimental conditions.

Moreover the evolution of the volume ratio with the friction coefficient presented on **Figure 23** seems to show that there is a limite value for volume ratio as rarefaction effects increases, showing that there is also a limite for the angle of attack to optimize the L/D ratio at high altitudes for a given geometrie. As presented in **Figure 24** the maximum L/D value decreases drastically with the increase of the friction coefficient which reflects the increase of the viscous effects.

7. Conclusions

This work focuses on the aerodynamic behavior of hypersonic gliders at high altitude. An experimental study of the behavior of a classic waverider has been conducted under supersonic and rarefied flow conditions for different values of the knudsen number. Measurements of drag and lift forces have been undertaken in the Marhy wind tunnel in supersonic operation for Mach 2 and Mach 4 numbers and equivalent pressures at altitudes between 5 km and 80 km. The experimental results showed that rarefaction effects produce a degradation of the flight performance of these waveriders, with a linear correlation between the value of the L/D ratio and the Knudsen number under the conditions studied. This is mainly due to the friction forces which increase with rarefaction due to viscous effects. This work also presents numerical simulation results obtained with the Bird DS3V code. The accomodation coefficients have been obtained with the CLL model for which the reflection and absorption coefficients have been adjusted according to the experimental conditions to be simulated. A linear correlation is also shown between these coefficients and the knudsen number. This study will be continued with an experimental study carried out in the marhy tunnel in hypersonic configuration at Mach 20 for an altitude of 100 km. First, the same waverider geometry will be studied, then a geometry will be optimized to improve the aerodynamic performances for more viscous flow conditions like those tested at Mach 20.

Nomenclature

λ	Mean free path (m)
μ	Dynamic viscosity (Pa.s)
ρ	Density (kg.m^{-3})

Θ	Angle incidence of test model (Degres)
Cd	Drag coefficient
CL	Lift coefficient
C_f	Friction coefficient
D_0	Wave drag
D_f	Friction drag
D_b	Base drag
Kn	Knudsen number
L/D	Lift-to-drag ratio
Ma	Mach number
p	Pressure (Pa)
Re	Reynolds number
$Re2$	Reynolds number after choc
R_m	Specific gas constant = 287.058 $m^2.s^{-2}.K^{-1}$
S	Horizontal prejected area
T	Temperature (K)
U	Flow speed ($m.s^{-1}$)
V	Internal volume

Subscript

∞	Free-stream conditions
0	Stagnation conditions
w	Wall conditions

References

[1] Ding, F., Liu, J., Shen, C. B., Liu, Z., Chen, S. H., & Fu, X. (2017). An overview of research on waverider design methodology. Acta Astronautica, 140, 190-205.

[2] Phoenix, A. A., Rogers, R. E., Maxwell, J. R., & Goodwin, G. B. (2018). Mach Five to Ten Morphing Waverider: Control Point Study. Journal of Aircraft, 1-12.

[3] Lewis, M. J., & McRonald, A. D. (1992). Design of hypersonic waveriders for aeroassisted interplanetary trajectories. Journal of Spacecraft and Rockets, 29(5), 653-660.

[4] Liu, W., Zhang, C. A., Han, H. Q., & Wang, F. M. (2016). Local piston theory with viscous correction and its application. AIAA Journal, 55(3), 942-954.

[5] Anderson, Jr, J. D. (2006). Hypersonic and high-temperature gas dynamics. American Institute of Aeronautics and Astronautics.

[6] Bowcutt, K. G., J. D. Anderson, and C. Diego. "Viscous optimized hypersonic waveriders. 1987." AIAA Paper: 87-0272.

[7] Rault, D. F. (1994). Aerodynamic characteristics of a hypersonic viscous optimized waverider at high altitudes. Journal of Spacecraft and Rockets, 31 (5), 719-727.

[8] Anderson, JR, John, Cang, J., & Mclaughlin, T. (1992, January). Hypersonic waveriders-Effects of chemically reacting flow and viscous interaction. In 30th Aerospace Sciences Meeting and Exhibit (p.302)

[9] Anderson, J. D., Lewis, M. J., Kothari, A. P., & Corda, S. (1991). Hypersonic waveriders for planetary atmospheres. Journal of Spacecraft and Rockets, 28(4), 401-410.

[10] Capriotti, D. (1987). Viscous optimized hypersonic waveriders. In 25th AIAA Aerospace Sciences Meeting (p. 272).

[11] Liu, W., Zhang, C. A., & Wang, F. M. (2018). Modification of hypersonic waveriders by vorticity-based boundary layer displacement thickness determination method. Aerospace Science and Technology, 75, 200-214.

[12] Rault, D., Willmoth, R., & Bird, G. (1991, June). An efficient DSMC algorithm applied to a delta wing. In 26th Thermophysics Conference (p. 1316).

[13] Kautz,F. A. II, , & Baron, J. (1991). Direct simulation of waveriders in hypersonic rarefied flow. In 26th Thermophysics Conference (p. 1317).

[14] Lewis, M. J., & Chauffour, M. L. (2005). Shock-based waverider design with pressure gradient corrections and computational simulations. Journal of aircraft, 42(5), 1350-1352.

[15] Lobbia, M. A., & Suzuki, K. (2014). Experimental investigation of a Mach 3.5 waverider designed using computational fluid dynamics. AIAA Journal, 53(6), 1590-1601.

[16] Nagashetty, K., Medhi, B., Sriram, R., Jagadeesh, G., & Reddy, K. P. J. (2017). Tomographic Visualization of the Hypersonic Flow Field over a Waverider. In 30th International Symposium on Shock Waves 2 (pp. 1437-1440). Springer, Cham.

[17] Rolim, Tiago Cavalcanti, et al. "Experimental results of a Mach 10 conical-flow derived waverider to 14-X hypersonic aerospace vehicle." Journal of Aerospace Technology and Management 3 (2011): 127-136.

[18] Rolim, T. C., Toro, P. G. D. P., Minucci, M. A. S., Oliveira, A. D. C. D.,

& Follador, R. D. C. (2011). Experimental results of a Mach 10 conical-flow derived waverider to 14-X hypersonic aerospace vehicle. Journal of Aerospace Technology and Management, 3(2), 127-136.

[19] Galloway, E. M., Gilmore, M., Jeffery, R., & Harvey, J. (1991). Heat transfer to a delta wing and two waverider wings in rarefied hypersonic flow. Rarefied gas dynamics, 505-513.

[20] Macrossan, M. N. (2006, July). Scaling parameters for hypersonic flow: correlation of sphere drag data. In Twenty-fifth Int. Symp. Rarefied Gas Dynamics, St. Petersburg, Russia (pp. 759-764).

[21] Orlik, E., Fedioun, I., & Davidenko, D. (2011). Boundary-layer transition on a hypersonic forebody: experiments and calculations. Journal of Spacecraft and Rockets, 48(4), 545-555.

[22] Fedioun, I., & Orlik, E. (2012). Boundary layer transition on the LEA hypersonic vehicle forebody. In 18th AIAA/3AF Int. Space Planes & Hypersonic Systems & Technol. Conf. (p. 5864).

[23] Rodi, P., & Bennett, G. (2012, June). High lift-to-drag ratio waveriders for missions in the Martian atmosphere. In 30th AIAA Applied Aerodynamics Conference (p. 3221).

[24] Knittel, J. (2015). Aero-Assisted Spacecraft Missions Using Hypersonic Waverider Aeroshells (Doctoral dissertation).

[25] Lunan, Duncan A. "Waverider, a revised chronology." 20th AIAA International Space Planes and Hypersonic Systems and Technologies Conference. 2015.

[26] Nonweiler, T. R. F. "Aerodynamic problems of manned space vehicles." The Aeronautical Journal 63.585 (1959): 521-528.

[27] Rasmussen, Maurice L. "Waverider configurations derived from inclined circular and elliptic cones. "Journal of Spacecraft and Rockets 17.6 (1980): 537-545.

[28] Ding, Feng, et al. "An overview of research on waverider design methodology." Acta Astronautica 140 (2017): 190-205.

[29] Anderson, JR, J., Frederick Ferguson, and MARK LEWIS. "Hypersonic waveriders for high altitude applications." 29th Aerospace Sciences Meeting. 1991.

[30] Townend, L. H. "Research and design for lifting reentry." Progress in Aerospace Sciences 19 (1979): 1-80.

[31] Rolim, T. C. (2009). Experimental analyisis of a hypersonic waverider (Doctoral dissertation, Instituto Tecnológico de Aeronáutica).

[32] Boutemedjet, Abdelwahid, et al. "Wind tunnel measurement of small values of rolling moment using six-component strain gauge balance." Measurement 116 (2018): 438-450.

[33] Reis, M. L. C. C., R. M. Castro, and O. A. F. Mello. "Calibration uncertainty estimation of a strain-gage external balance." Measurement 46.1 (2013): 24-33.

[34] Aroesty, Jerome. Sphere drag in a low density supersonic flow. California Univ. Berkley Inst. of Engineering Research, 1962.

[35] Hoffmann, Karl. An introduction to measurements using strain gages. No. BOOK. Darmstadt: Hottinger Baldwin Messtechnik, 1989.

[36] Higson, GRt. "Recent advances in strain gauges." Journal of Scientific Instruments 41.7 (1964): 405.

[37] Tufte, O. N., and D. Long. "Recent developments in semiconductor

piezoresistive devices." Solid-state electronics 6.4 (1963): 323-338.

[38] Schlegat, Thomas, and Klaus Hannemann. "Experimental Investigation of Rarefaction Effects on Aerodynamic Coefficients of Slender and Blunt Re-entry Vehicles." International Conference on High-Speed Vehicle Science and Technology 2018 (HiSST). 2018.

[39] Coumar, Sandra. Study of physical mechanisms induced by a plasma actuator for super/hypersonic rarefied flows applied to atmospheric entries. Diss. Université d'Orléans, 2017.

[40] Boyd, Iain D. "Predicting breakdown of the continuum equations under rarefied flow conditions." AIP Conference Proceedings. Vol. 663. No. 1. American Institute of Physics, 2003.

[41] Riabov, Vladimir V. "Comparative similarity analysis of hypersonic rarefied gas flows near simple-shape bodies." Journal of Spacecraft and Rockets 35.4 (1998): 424-433.

[42] Probstein, Ronald F., and Nelson H. Kemp. "Viscous aerodynamic characteristics in hypersonic rarefied gas flow." Journal of the Aerospace Sciences 27.3 (1960): 174-192.

[43] Bird, G. A. "Molecular gas dynamics and direct simulation Monte Carlo." Clarendon Press Oxford Science (1998).

[44] Bird, G. A., and D. S. M. C. Visual. "Program for Three-Dimensional Flows." The DS3V Program User's Guide, Version 2 (2006).

[45] Zuppardi, Gennaro, and Giuseppe Mongelluzzo. "Computer analysis of rarefied aerodynamics around a winged space-plane for Mars entry." Advances in Aircraft and Spacecraft Science 8.2 (2021): 169.

[46] Zuppardi, Gennaro. "Influence of partial accommodation coefficients on the aerodynamic parameters of an airfoil in hypersonic, rarefied flow." Advances in aircraft and spacecraft science 2.4 (2015): 427.

[47] Hedahl, Marc O., and Richard G. Wilmoth. Comparisons of the Maxwell and CLL gas/surface interaction models using DSMC. National Aeronautics and Space Administration, Langley Research Center, 1995.

[48] Rasmussen, M. "Viscous effects on the performance of cone-derived waveriders." 10th Atmospheric Flight Mechanics Conference. 1983.

Chapter 4

Plasma Preheating Technology for Ablation Studies of Hypersonic Reentry Vehicles

Abstract

Measurements of mass ablation rates in hypersonic flows are used to calibrate computational models. A novel plasma technique for preheating axisymmetric samples of heatshield materials has been developed and applied to ablation of graphite sample through this work. The experimental probe was very similar to the European standard probe, normal to the flow. The Scanning Electron Microscope (SEM) was used to examine the surface characteristics; and the experiments show a significant spatial variation in thickness loss for the graphite test material over the disc radius though the spatial variation was still largely axisymmetric. The present work does support reasonable contributions to reentries and can further be developed to validate computational models under conditions that replicate characteristics of reentry flights.

Keywords: Heatshield, Plasma, Preheating, Technology, Ablation, Hypersonic, Planetary, Re-entry vehicles, Surface temperatures

1. Introduction

To achieve practical reentries of flight vehicles, speeds in excess of the earth's escape velocity ($\geq$ 11 km/s) are needed [1]. On reentry, this hypersonic speed is slowed down by viscous drag: firstly, thermal energy is created through the deceleration and compression of the high velocity incoming flow to a high pressure and high temperature in the hypersonic boundary layer close to the surface (heatshield); and secondly the convective and radiative heating associated with high temperatures accelerates the wall reactions [2]. The blunt shape of heatshields consisting of carbon-based materials [3] are effective structures that enable most of the generated heat to be carried away from the vehicle [4]. Computational models of the heat loads [5, 6] experienced during atmospheric reentries are continually being updated [7, 8]. The validation of these models is critical for the safe and economic design of future flight vehicles [9]. Proper analysis of heat flux [10], gas/surface interactions [11], and properties of heatshield materials [12] are needed for ablation performance and evaluations [13, 14]. Various thermochemical processes during reentry [15], descent, and landing also support the ablation of heatshields [16]. Flow parameters [17] such as enthalpy, stagnation pressure, velocity and heat flux are identified in **Figure 1**. It explains how the velocity of a reentry Stardust is slowed from 12.8 km/s to almost 2 km/s. A significant reduction in velocity is experienced

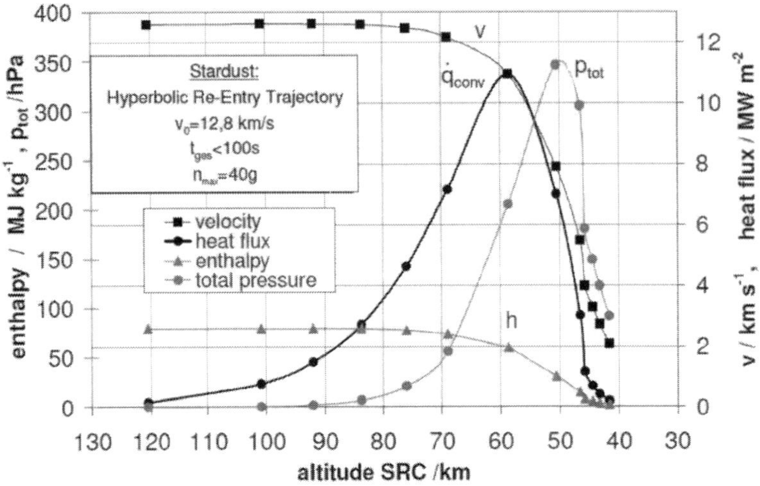

Figure 1.
Stardust trajectory parameter [18].

between 70 to 40 km altitude due to higher atmospheric density resulting in more effective drag. At the same time, the heat-flux rises to a peak of 11 MW/m^2 at 62 km altitude. All space vehicles like Apollo, Huygens, Stardust, Hayabusa and Orion generate heat at reentry as a consequence of the drag used to reduce speed [19].

2. Overview

This work helps to explain quantitative measurements of graphite ablation and oxidation rates in a hypersonic boundary layer using a new hot-surface plasma preheating technique to raise the temperature of the model to the required surface temperature for reentry studies in a hypersonic impulse facility [20]. Hot-wall reentry tests have been carried out in hypersonic impulse facilities [21] with temperatures characteristic of ablators in hypervelocity reentry of approximately 2000–3000 K in 8.6 km/s for Earth reentry flow [22]. Experiments with electrically preheated graphite samples using uniform width profiles were conducted at the University of Queensland's X2 expansion tunnel [23] for surface temperatures from 1770 to 2410 K used to target the carbon–nitrogen violet band [24]. All of these used resistive heating on a uniform width profile and not a plasma preheating technique. Recent experiments based on Orion reentry conditions used a surface temperature of about 2800 K [25] and Apollo 4 lunar return speed of 11 km/s reportedly experienced a surface temperature of about 2400 K [26], while typical reentry surface temperatures for Space Shuttles was about 1740 K [25]. This present work has achieved temperatures in excess of 2500 K on a preheated graphite surface used for the assessment of mass loss through ablation in a Mach 4.5 and Mach 6 flows. The use of a plasma as the means of heating the disc was proven to be effective for reaching temperatures of these magnitudes [27] and the process has potential to be more widely used for similar experiments [28]. The technique uses a plasma heating source with argon flow [29]. The work reports the first-time quantification of ablation and oxidation rates of a heated carbon disc up to 500 milliseconds duration in 500 Pa vacuum pressure in a hypersonic wind tunnel facility using a plasma preheating methodology.

3. Plasma preheating methodology

The model in the present work can be likened to the Euro-model of the European Space Agency (ESA) which is a flat-faced cylinder of 50 mm in diameter with rounded corners of 1 mm radius, positioned at a 0° angle of attack to the incoming flow. Preheating of the graphite disc was achieved with plasma [30, 31] generated by a DC current [32, 33] between a tungsten electrode and the back (downstream) side of the disc. **Figure 2** is a sectional view of the model which illustrates the heat transfer processes from the hot plasma to the disc. The orientation of the tungsten inert gas (TIG) is centralised in the model to enable even thermal spread from the centre to the edges. This made it possible for the probe model to assume an axisymmetric orientation.

The disc was set up and heated for 15 seconds. The flow was initiated after 15 seconds of heating, manually timed. The heating remained on during the flow and for about 0.5 seconds after the flow stopped. The present work adopts heatshields of a constant thickness of 2 mm and the thickness was considered suitable for ablation experiments [27]. Using graphite material and a current rating of 400 A, it took about 15 seconds for the plain disk-sample to attain a steady state. Using this current and invoking graphite material properties into FEA simulation with Ansys [29], the results from simulations were found to have a good agreement with experiments as shown in **Figure 3**. The first 15 seconds was for heating, followed by the flow which lasted for about 0.5 seconds, and then finally cooling due to power cut-off from supply. The Schlieren technique based on the principle of changing densities in the gas was used to identify the establishment of the bow shock and therefore the commencement of flow [35]. Obtaining the plenum pressure from pressure survey array, run-times were matched with Schlieren images to identify when flow starts [27]. The high-speed camera was set at a frame-rate of 2500 fps and frames at the specific points during the flow are shown in **Figure 3**.

Figure 4 illustrates the flow-dynamics and the associated aerothermodynamic gradient along the surface, where the source of heat flux for the heatshield is the plasma inside the probe. The temperature is driven by the plasma at the backside of experimental sample, while the aerodynamic flow is driven by the forced convection from hypersonic impulse facility at the frontside of sample (heatshield specimen). The aerodynamic flow-velocity at the surface of the experimental

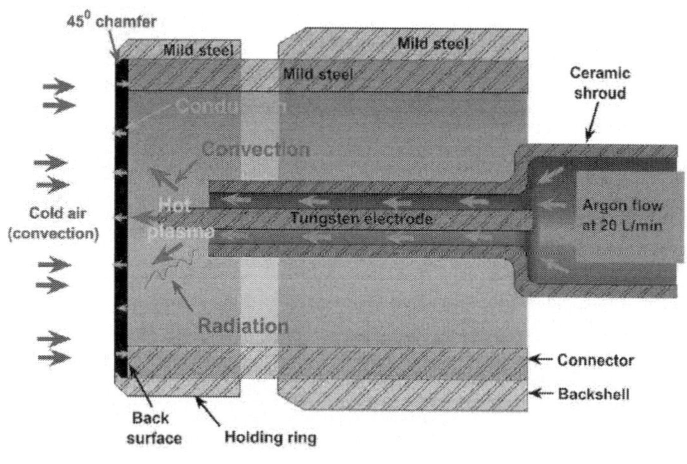

Figure 2.
The new plasma preheating technique illustrating the heat transfer processes from hot plasma to graphite disc [34].

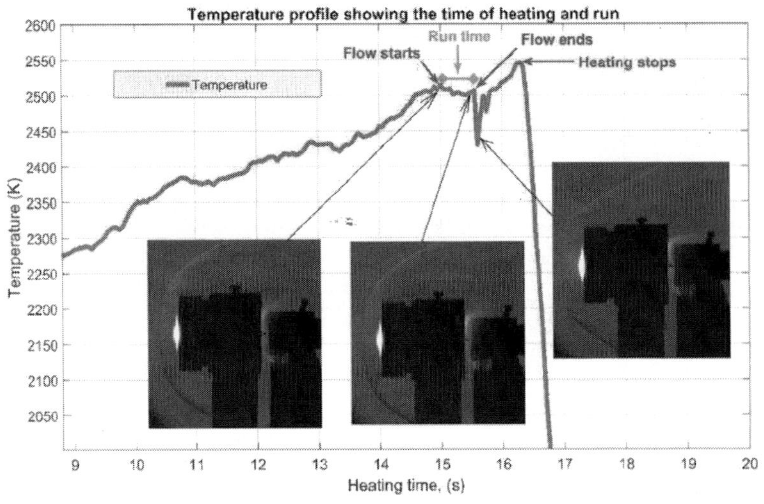

Figure 3.
Transient behaviour of stagnation surface temperatures with heatshields of 2 mm thickness [28].

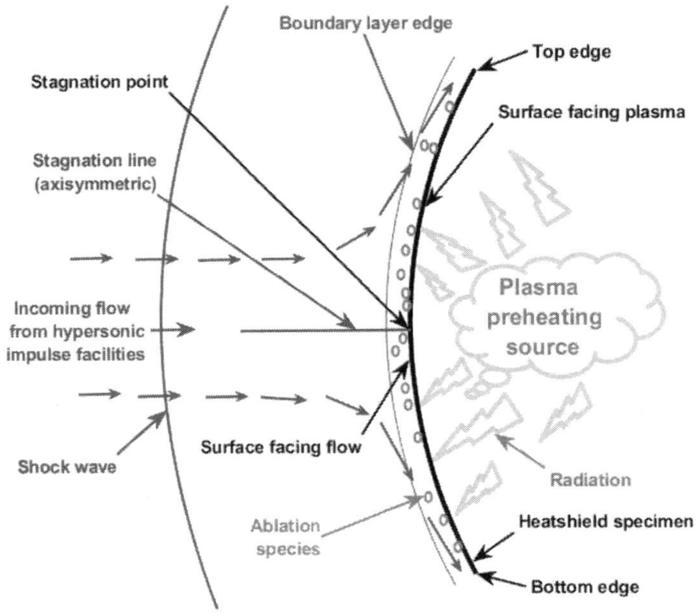

Figure 4.
Schematic illustration of surface aerothermodynamic flow properties [36].

sample increases from the stagnation point to the edges, while the surface temperature decreases from the stagnation point to the edges. While the aerothermodynamic flow (cooling of sample) occurs at the front, the plasmadynamic flow (heating of sample) occurs at the backside of the disk. The plasma zone describes the region occupied by the plasma [36]. At steady state conditions, the inert gas flows through the shroud, gain some heat energy from the centralised hot tungsten electrode, then experiences a drastic rise in enthalpy as it passes through the hot plasma towards the heatshield, before finally exiting via the vent as shown in **Figure 5**.

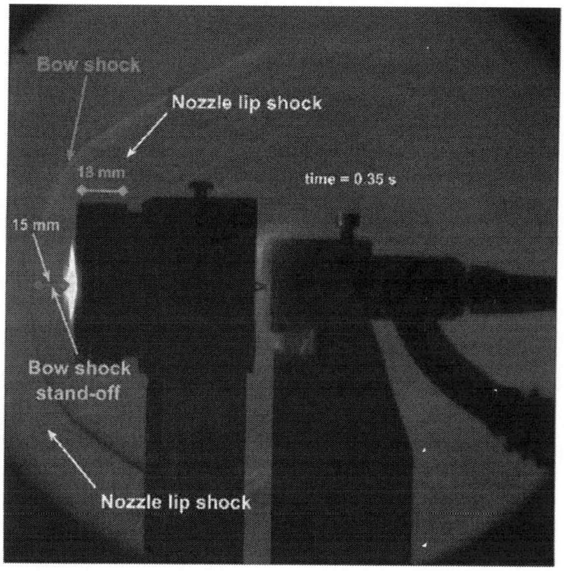

Figure 5.
Test conducted at Mach 4.5 to validate plasma preheating technology [28].

The aerothermodynamic flow properties for the plain disk of radius 25 mm were simulated using Ansys Fluent CFD. The CFD results using an axisymmetric graphite sample, showed a good temperature distribution for ablation rate experiments in hypersonic impulse facilities. The temperature is driven by the plasma at the backside of experimental sample, while the flow-field is driven by the hypersonic impulse facility at the frontside of sample [36]. **Figure 6** shows the density variations from Mach 0 to 6, using 2D axisymmetric simulations. The density gradually increased from no-flow conditions to Mach 6 hypersonic flow. Numerical and experimental analysis at Mach 4.5 have been presented extensively by the author in other publications [28, 29].

The reaction species were simulated by using 21% O_2 and 79% N_2 at the hypersonic inlet, to flow over heated graphite surface [29]. The pressure gradient in **Figure 7a** shows the bow shock standoff distance, while **Figure 7b** shows the contour result of Carbon II oxide species using 2D axisymmetric simulation in Mach 4.5 flow. The ablation rate gradually decreased from stagnation point to the edges of the disk. Along the stagnation line of **Figure 7a**, the upstream edge of the experimental bow shock was about 15 mm from the surface, very similar to the bow shock position described in **Figure 5**. The numerical and experimental results from Mach 4.5 were very similar to that of Mach 6.

Flow properties associated with the present work were obtained from CFD simulations using ANSYS Fluent. **Figure 8** shows the simulation results using a density-based solver for the physical parameters along the stagnation line for a plain disk sample [29]. Surface temperatures and boundary conditions in the CFD were set using measurement information from experiments. **Figure 8** indicates that the temperature has not changed significantly until within 1 mm from the wall, so the concentration of products was not entirely driven by temperature alone. The Mach 4.5 flow parameters were compared with that of Mach 6. The values along the stagnation line were roughly the same. The origin of the horizontal-axes in these figures is at the surface of the disk. In the case of Mach 4.5, the pressure increased from 400 Pa to 18.6 kPa across the shock in the flow direction. The gas density increases from 0.045 kgm^3 to 0.22 kgm^3 across the shock, and continues to rise to a

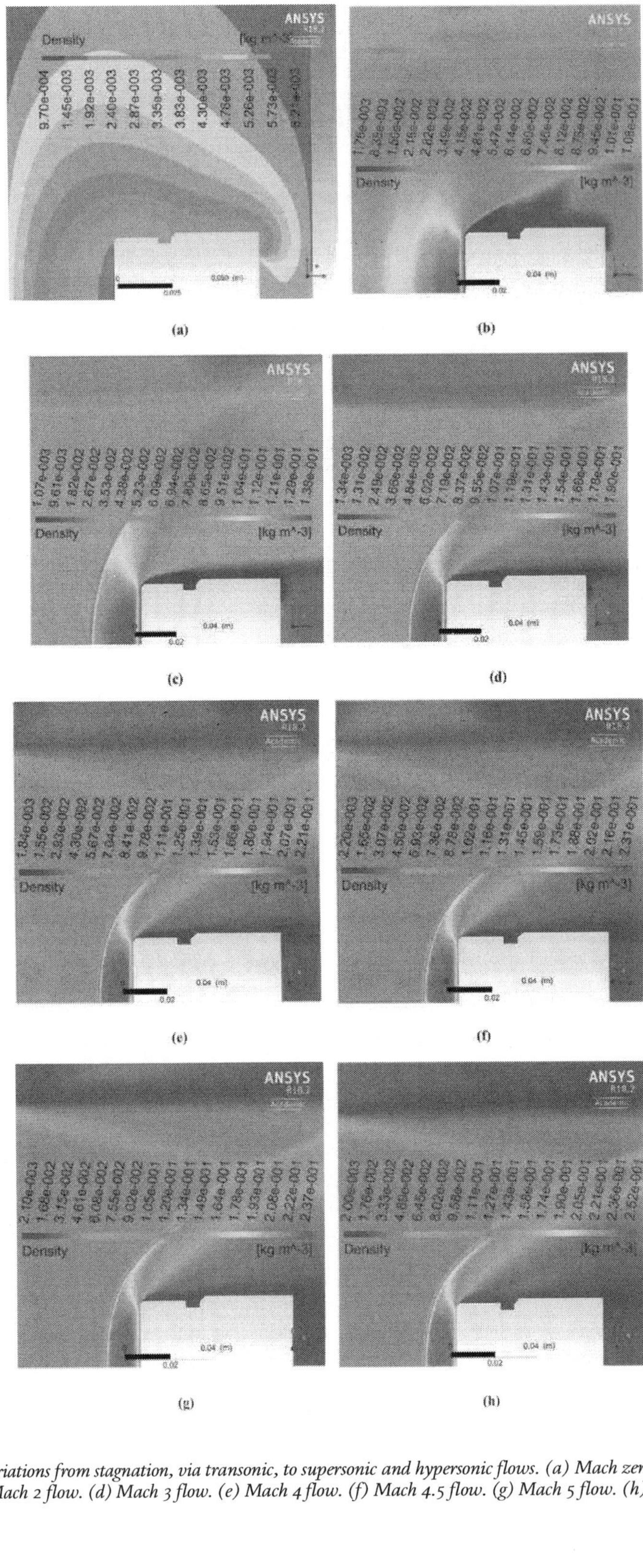

Figure 6.
Density variations from stagnation, via transonic, to supersonic and hypersonic flows. (a) Mach zero. (b) Mach 1 flow. (c) Mach 2 flow. (d) Mach 3 flow. (e) Mach 4 flow. (f) Mach 4.5 flow. (g) Mach 5 flow. (h) Mach 6 flow.

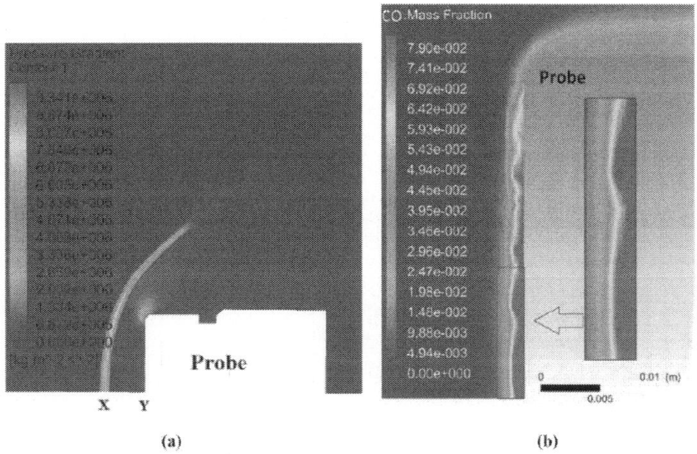

Figure 7.
Flow and species parameters at Mach 4.5 flow from simulations. (a) Bow shock standoff distance. (b) CO mass fraction [29].

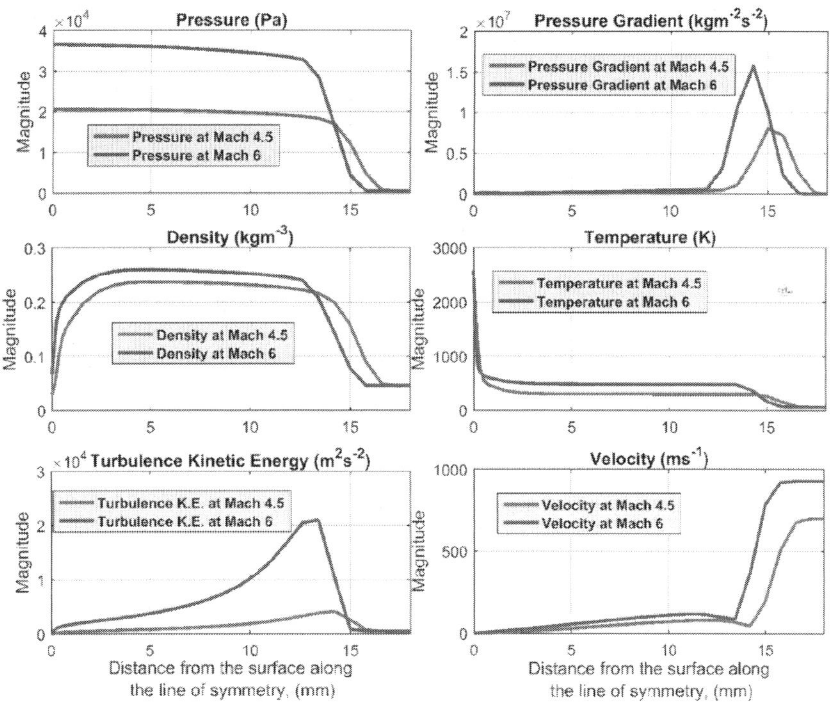

Figure 8.
Boundary layer flow parameters from simulations at Mach 4.5 and Mach 6.

maximum value of 0.23 kgm^3 at about 4 mm from the wall; and finally dropping to a minimum value of 0.033 kgm^3 at the wall. The static temperature rises from 60 K to 272 K across the shock, maintaining this relatively constant value to about 4 mm from the hot surface, then rises to a maximum value of about 2500 K at the wall, which explains the density change [34]. The turbulence kinetic energy is greatest at the shock and falls to almost zero at the stagnation point, suggesting a laminar boundary layer. The gas velocity also decreased across the shock from about 700 m/s to about 51 m/s, and then experiencing a small rise immediately after the shock and gradually falling to zero at the stagnation point.

No changes in the gas properties were evident until approximately 4 mm from the wall as shown in **Figure 9**. The mass fraction of N_2 behaves in a similar way to that of O_2 with a sharp drop near the surface as a result of the increase in mass fraction of reaction products entering the flow from the surface. The mass fraction of molecular oxygen dropped from 21% to about 8% at the wall while that of molecular nitrogen dropped from 79% to about 67% at the wall. Neither the molecular oxygen nor nitrogen concentration dropped to zero at the wall as the temperature reached by the gas was not sufficient for complete dissociation. This indicates that a mixture of N_2 and O_2 was still present in the reacting boundary layer. A contribution to the reduced concentration of molecular oxygen results from its consumption in combustion. The molecular oxygen was only dissociated in very close proximity to the surface as the gas temperature only reached the dissociation temperature within 0.0025 mm from the wall. Species transports are driven by flow properties. Thermally initiated chemical reactions are the formation process of all carbonaceous species in **Figure 9** except for the carbon sublimation species, C, C_2, C_3. The sublimation species C, C_2, and C_3 were almost zero. The CO_2 species are formed from further oxidation of CO species, thus making CO_2 a secondary reaction. The result also shows that CO_2 has a sharp drop from the peak value to almost zero at the surface within the experimental temperature limit of 2530 K. The CN species distribution also show a similar pattern to that of CO_2. This also suggests that CN is not a product of direct surface reaction within the experimental temperature limit. This further supports the absence of dissociated nitrogen atoms which would otherwise aid direct formation of CN at the surface [29]. The simulations did

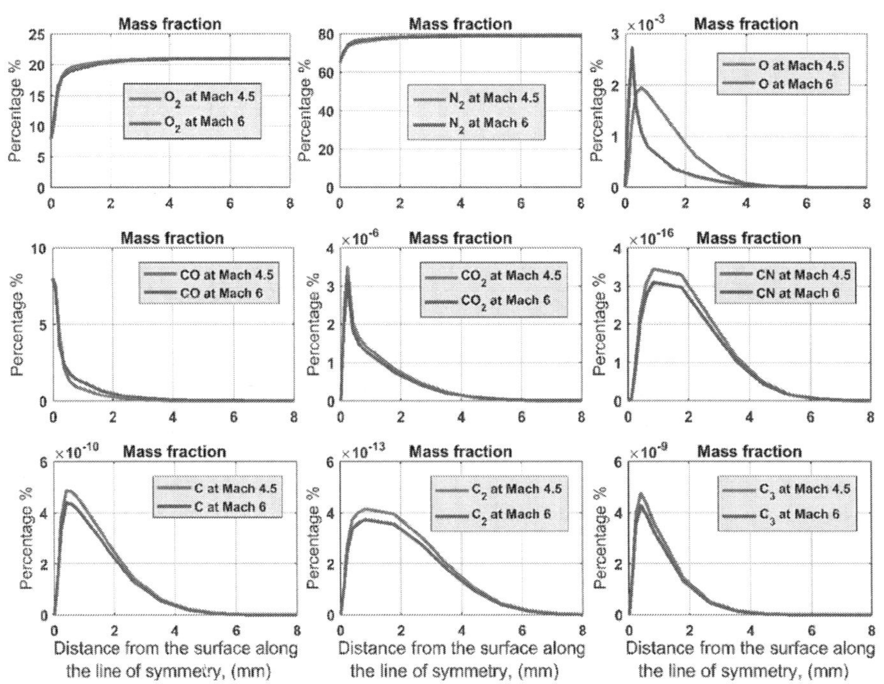

Figure 9.
Boundary layer species concentrations from simulations at Mach 4.5 and Mach 6.

not predict atomic nitrogen in the present work. This is because the temperature needed to cause N_2 dissociation was not achieved in the present work. The CO formation was the major contributor to graphite mass loss and contributions from all other carbonaceous species were insignificant [29]. All carbonaceous species have their peak values around the stagnation region at the wall. The mass fraction of CO species rises to a maximum value at the wall. Mass fraction of CO_2 increases to a maximum near the wall and then drops closer to the wall. Simulation results from CFD show that the peak mass fraction of CO at the surface was about 7.9%. The CO_2 mass fraction was about six orders of magnitude lower than that of CO.

The information contained in this material is a new experimental method for heating material samples using a high temperature plasma arc fixture. This new fixture can be integrated within cold hypersonic flow wind tunnels in order to measure material ablation and mass loss due to aerodynamic flow effects. The mass loss measurements and microscopic images can be obtained to quantitatively and qualitatively observe material mass loss and surface modifications. Finite rate ablation chemistry has been used in the present work along with volumetric and wall surface reactions [29]. The technique makes it possible to document mass loss from their specimens using dimensional values. This invention focuses on ground-based characterisation techniques for thermal protection material analysis, thus making it a valuable tool for aerothermodynamics of reentry studies.

Well prepared experiments with clear scope and structure will be of importance to the hypersonic research community, especially in the field of new ground-testing facilities for ablation analysis. The presented technology is useful for detailed material characterisation, for example, in the context of material response model validation. It is a new test facility, which can adopt a new measurement technique to generate data for model calibration and validation. The technique is able to accurately replicate the hypersonic flow characteristics and heatshield conditions at reentries. This technology is indispensable when it comes to presenting a new experimental technique for ablation analysis, highlighting new techniques for ablation measurements and providing experimental data for code validation. The key contributions to reentries include: (1) the development of the new test technique, (2) the quantifiable measurements of mass loss, and (3) connecting the measurements made to ablation theory or models.

4. Surface ablation using plasma preheating technology

4.1 Theoretical approach

To effectively couple the solid and the flow-field during ablation, the mass, momentum, energy, and species have to be conserved. The surface mass balance for each species is given in Eq. (1), where $\dot{m}'_s$ is the mass flux of species s per second determined from the surface thermochemistry, ρ_s is species density in kg/m^3, $\vec{v}$ is the velocity vector representing the mass-averaged velocity leaving the surface, $\hat{n}$ is the unit normal vector to the surface but away from the wall, and the last term is the diffusion of species to or from the surface, where D is the diffusion coefficient in m^2/s, C_s is the species mass fraction [37].

$$\dot{m}'_s = \rho_s \vec{v}.\hat{n} - \rho D \vec{\nabla} C_s \hat{n} \qquad (1)$$

The surface energy balance is expressed in Eq. (2), where $\vec{q}_w$ contains both the heat conduction and the diffusive chemical heat flux, the first term on the right is the heat flux conducting energy away from the surface into the body, the second term is the radiation of heat from the surface into the flow, and the last term is the removal of energy from the surface due to mass removal [38].

$$\vec{q}_w = -K_{solid}\vec{\nabla}T_{solid}.\hat{n} - \varepsilon\sigma(T_w - T_a)^4 - \dot{m}'_w h_{o,w} \tag{2}$$

Generally, K_{solid} is the thermal conductivity of graphite sample in W/(m.K), T_{solid} is the temperature of graphite at the edges, ε is graphite emissivity, σ is the Stefan-Boltzmann constant in W/(m^2.K^4), T_w is the surface temperature, T_a is the surrounding temperature, and $h_{o,w}$ is species enthalpy in J/kg [39]. Unlike the negative heat flux to the wall from the flow in high enthalpy facilities, the present work adopts a positive heat flux from the wall to the flow.

4.2 Surface ccharacteristics at Mach 4.5 flow using the micrometre gauge and SEM

The micrometre measurements also recorded losses in thickness over successive runs. Four points were used across the disc at 0 mm, 8 mm, 16 mm, and 24 mm from the centre as shown in **Figure 10**. The measured thickness across the four

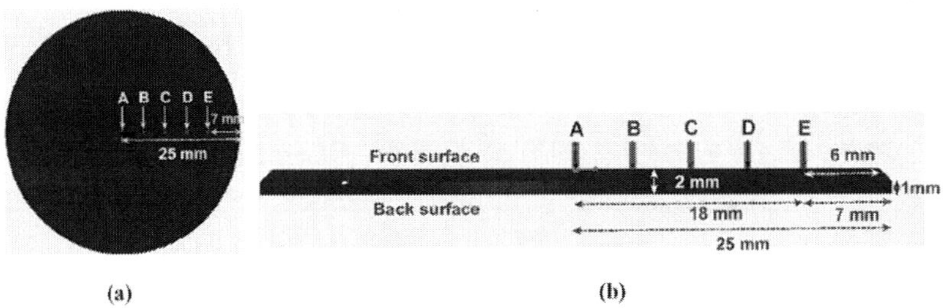

Figure 10.
Scanning locations on graphite surface. (a) Graphite specimen (top view). (b) Graphite specimen (side view not to scale).

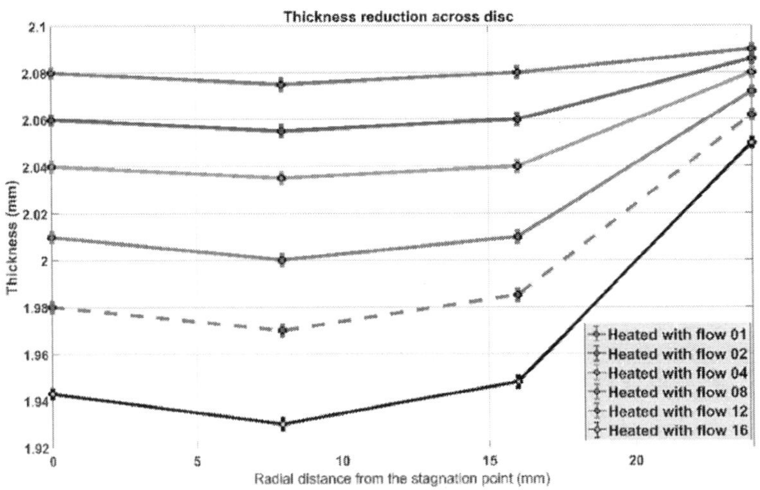

Figure 11.
Variation in thickness across disc along paths perpendicular to the surface at various radial positions using the micrometre gauge after 1, 2, 4, 8, 12 and 16 heated-with-flow runs.

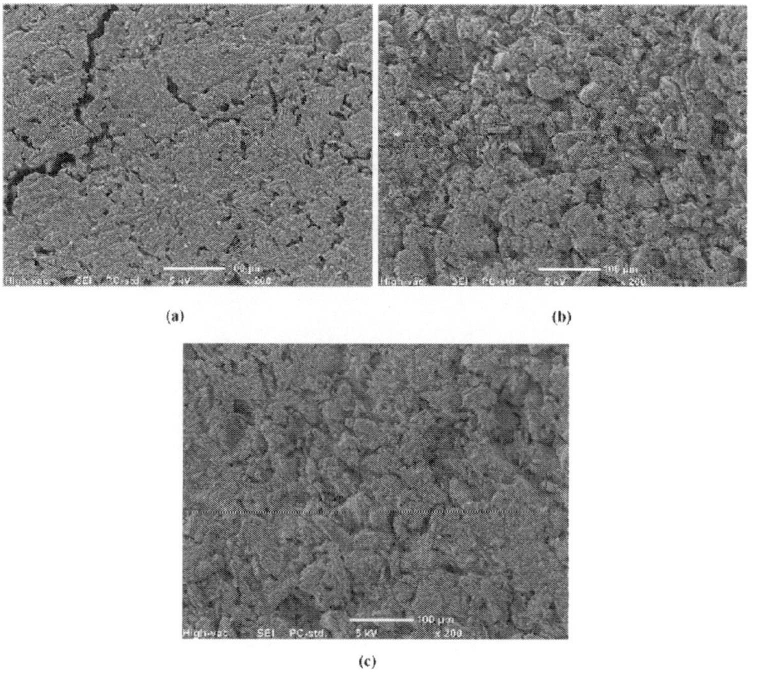

Figure 12.
SEM images after the second heated with flow run in proximity of points A, C and E. (a) In the proximity of point A. (b) In the proximity of point C. (c) In the proximity of point E.

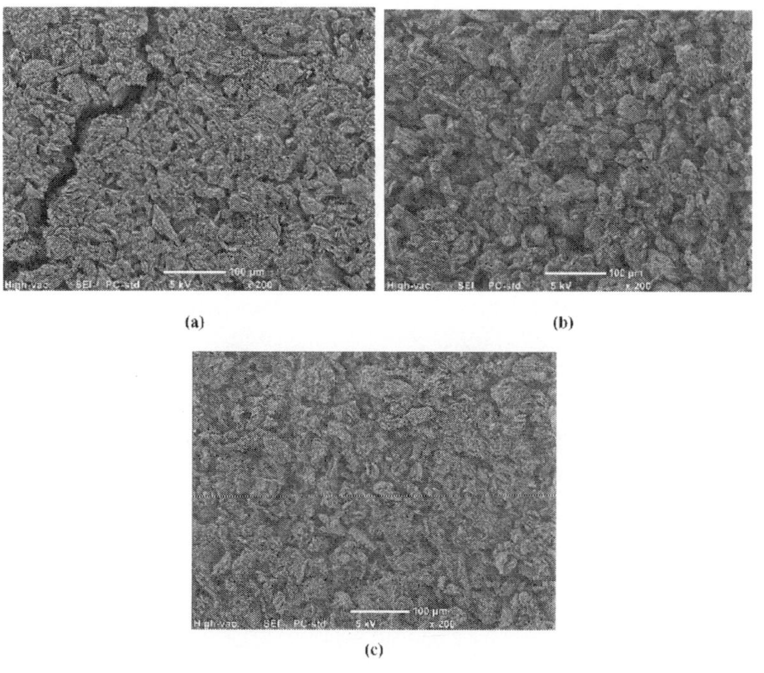

Figure 13.
SEM images after the fourth heated with flow run in proximity of points A, C and E. (a) In the proximity of point A. (b) In the proximity of point C. (c) In the proximity of point E.

locations is shown in **Figure 11**. The anvil and spindle of the micrometre (where the micrometre contacted the disc) were 6.5 mm in diameter. The disc deformation and the ablating surface meant that the contact from the micrometre was unlikely to accurately measure the local thickness at each point, but more likely measured a general thickness at the location in the vicinity of each point. **Figure 11** shows the general trend resulting from the ablation of the disc. The thickness from the micrometre measurements in **Figure 11** shows a general agreement with that obtained from the measuring arm [34]. The laser sheet visualisation confidently identifies that the material loss is not at a consistent rate along a radius of the disc, but that the trend of the material loss rate is relatively consistent over the duration

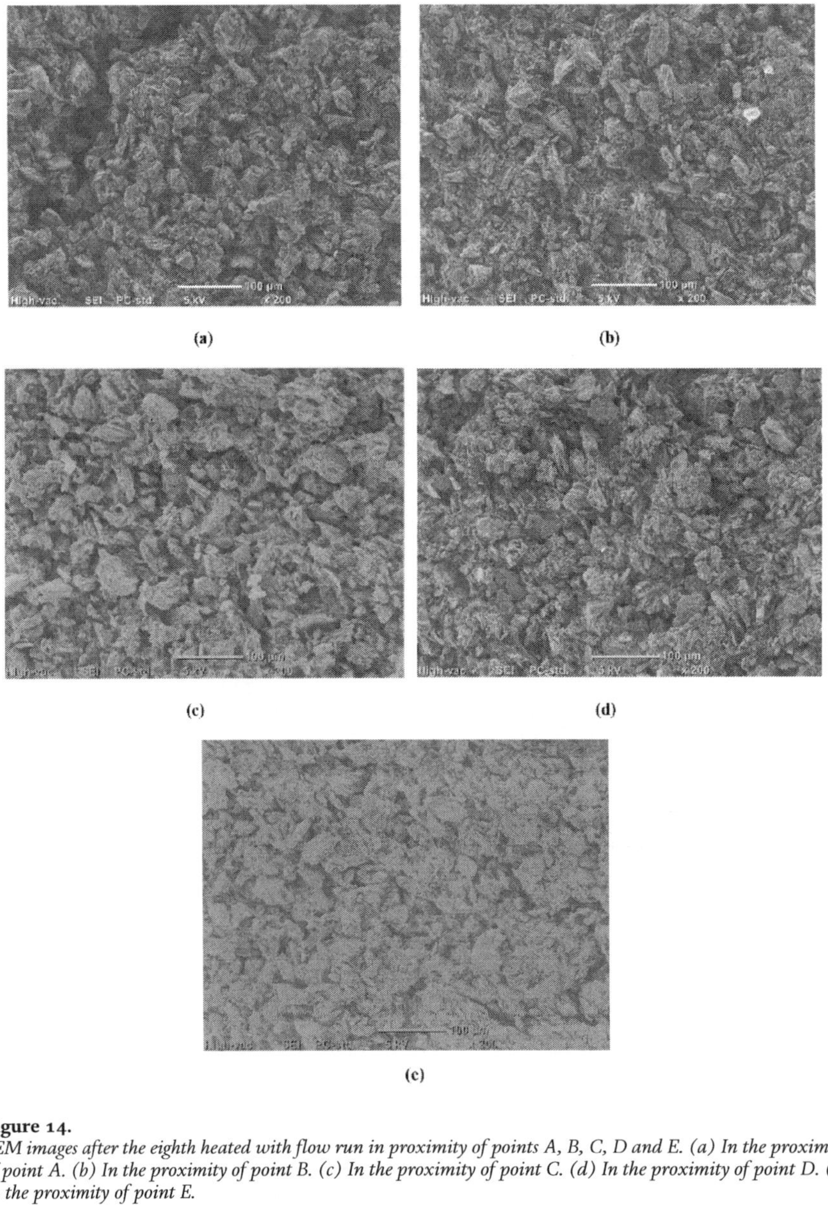

Figure 14.
SEM images after the eighth heated with flow run in proximity of points A, B, C, D and E. (a) In the proximity of point A. (b) In the proximity of point B. (c) In the proximity of point C. (d) In the proximity of point D. (e) In the proximity of point E.

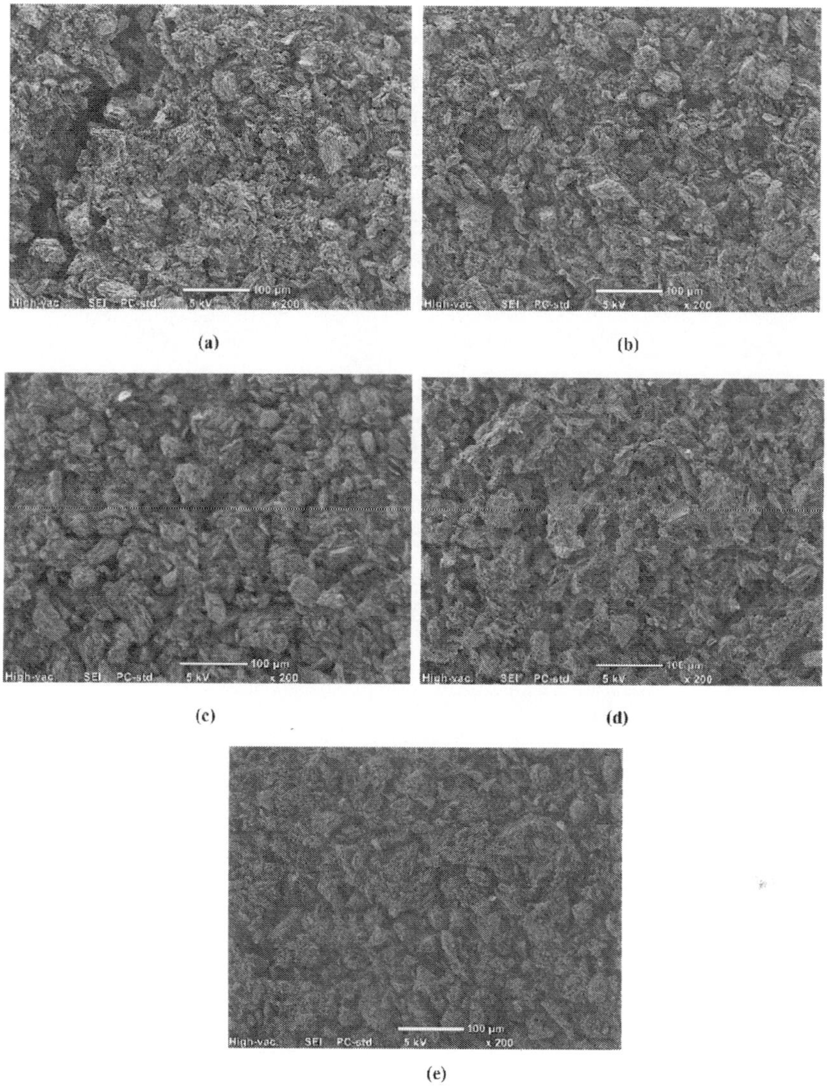

Figure 15.
SEM images after the twelfth heated with flow run in proximity of points A, B, C, D and E. (a) In the proximity of point A. (b) In the proximity of point B. (c) In the proximity of point C. (d) In the proximity of point D. (e) In the proximity of point E.

of the experiments. The experiments show a significant spatial variation in thickness loss for the graphite test material over the disc radius though the spatial variation was still largely axisymmetric.

Tools such as the Scanning Electron Microscope (SEM) was used for surface characteristics and the results from microscopy are presented in **Figures 12–18**. Five positions in total were chosen at 4.5 mm increments, shown as A, B, C, D, and E in **Figure 10**. This resulted in position E being 7 mm from the outside edge of the disc (6 mm from the chamfer used to retain the disc). During experiments, the actual points scanned in the SEM were not exactly the same; each scan was a representative area in close proximity to the points described by A, B, C, D and E.

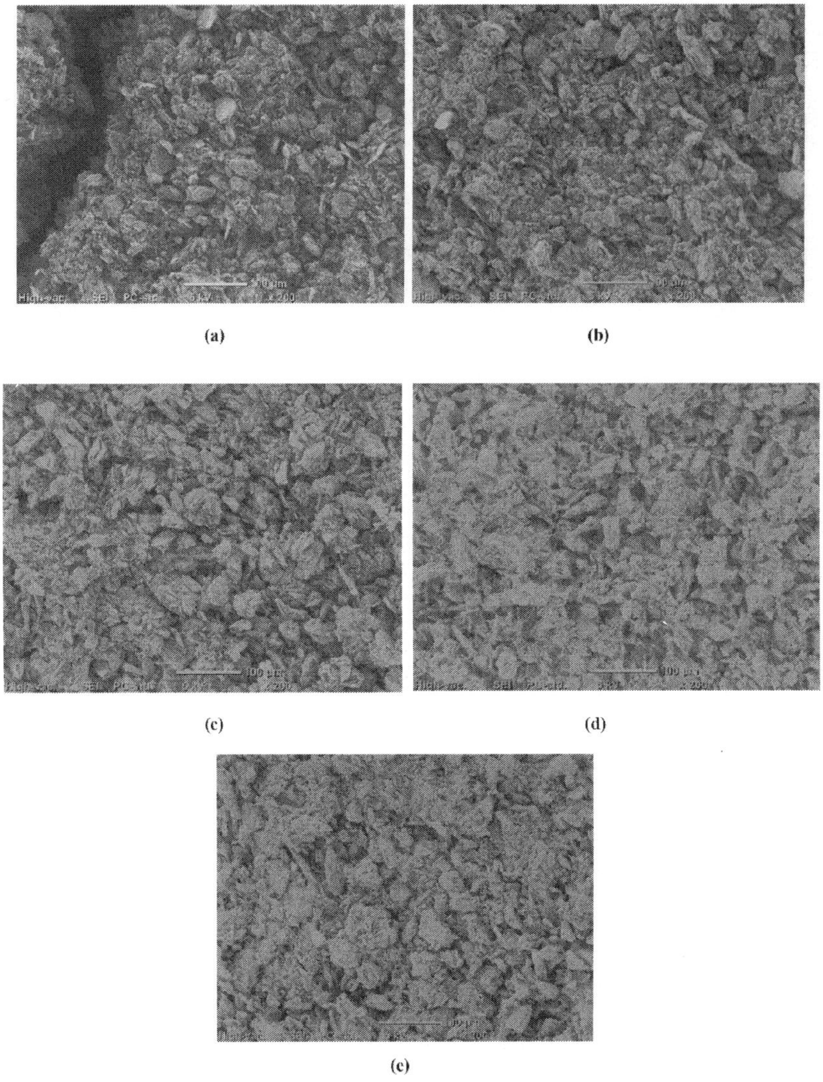

Figure 16.
SEM images after the sixteenth heated with flow run in proximity of points A, B, C, D and E. Back of disc. (a) In the proximity of point A. (b) In the proximity of point B. (c) In the proximity of point C. (d) In the proximity of point D. (e) In the proximity of point E.

5. Future work

The future work has optimised the aerothermodynamic efficiency for future experimental tests and can now generate stagnation point temperatures in excess of 3000 K. The two major improvements that are associated with the Next Generation Experimental Model (NGEM) include: (a) reducing the heat losses by conduction on heatshield sample by incorporating a thermal barrier in-between the test-sample and the backshell; and (b) incorporating six degrees of freedom in order to account for the variable angle of attacks for better manoeuvrability during reentry, descent and landing [36]. These modifications enable the NGEM to be smarter and more

Back of Disc

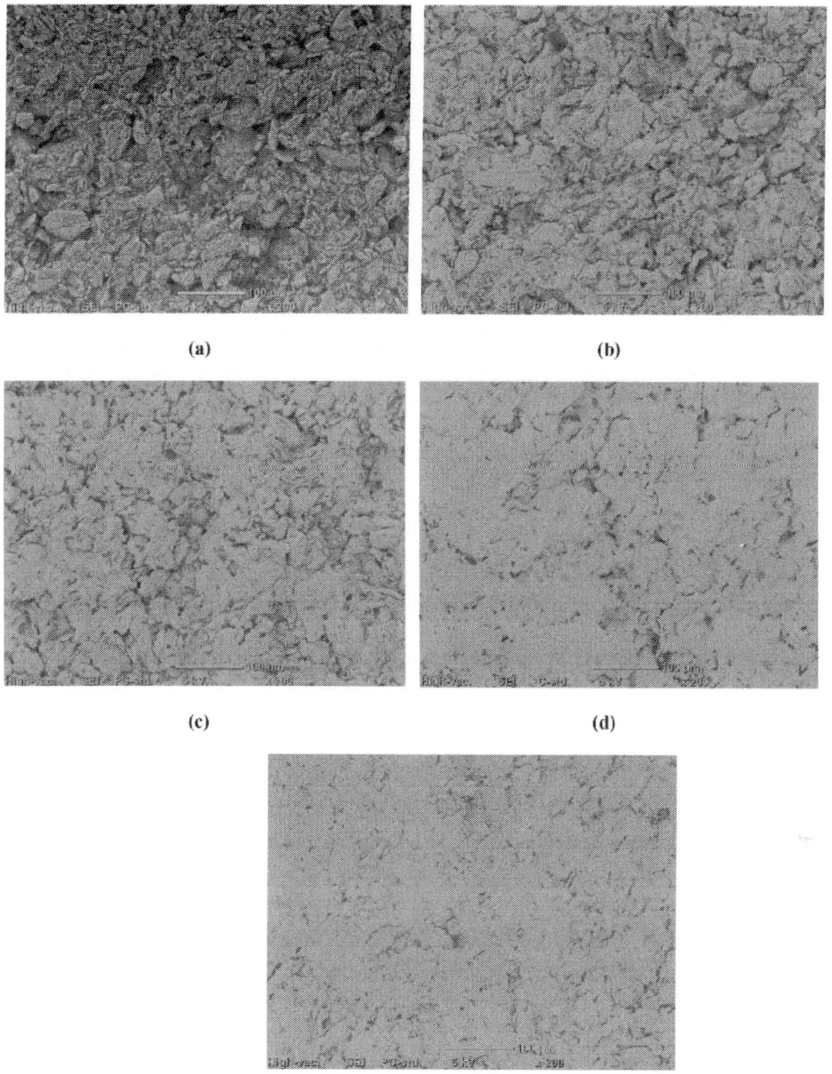

(a) (b) (c) (d) (e)

Figure 17.
SEM images after the twelfth heated with flow run in proximity of points A, B, C, D and E (back). (a) In the proximity of point A. (b) In the proximity of point B. (c) In the proximity of point C. (d) In the proximity of point D. (e) In the proximity of point E.

practically replicates real-flight vehicles as shown in **Figure 19**. The future experimental model has been fully developed for series of ablation tests in any reliable aerospace laboratory. Future experimental tests will encourage the use of spectroscopic measurements of ablation species and spatial microstructural studies using X-ray Microtomography. Infrared pyrometers and thermo-cameras are also needed to adequately monitor the surface temperature profiles across experimental samples. The next generation experimental model is not only expected to generate a stagnation point temperature of about 3200 K under the same experimental

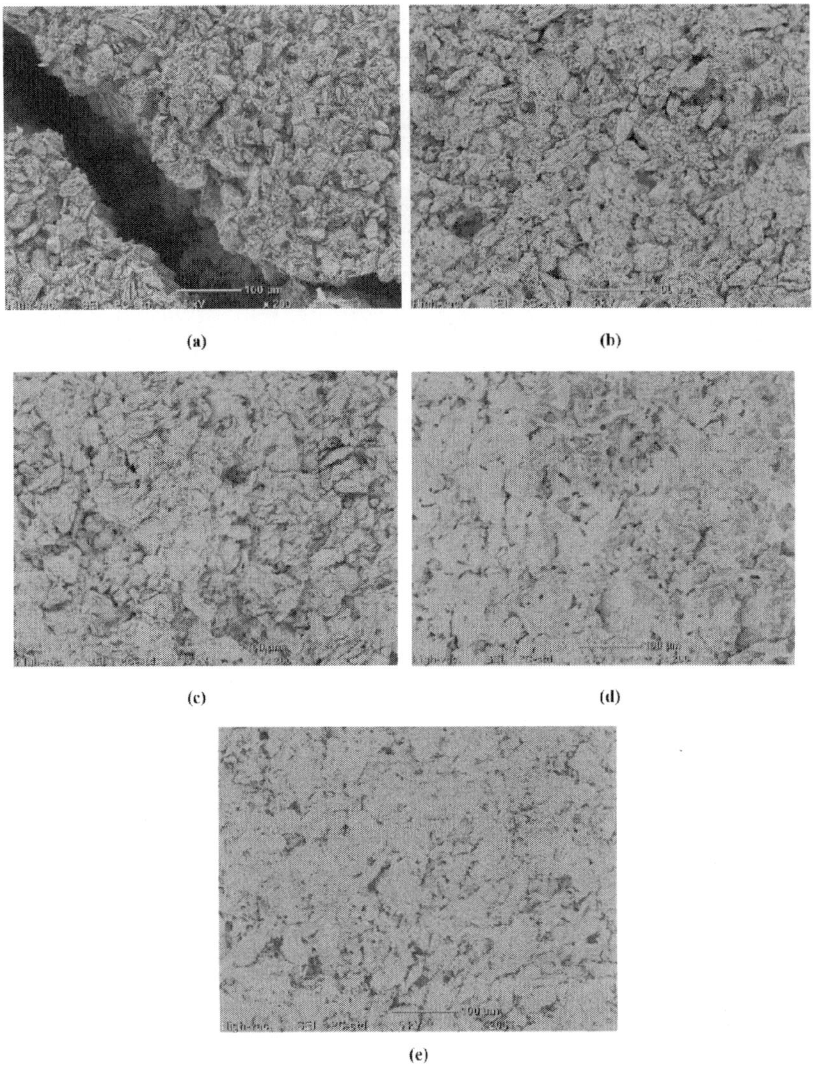

Figure 18.
SEM images after sixteenth heated-with-flow run in proximity of points A, B, C, D and E (back). (a) In the proximity of point A. (b) In the proximity of point B. (c) In the proximity of point C. (d) In the proximity of point D. (e) In the proximity of point E.

conditions, but also enables variable angle of attack for better manoeuvrability during reentry, descent and landing. These inclusions have never been attempted anywhere else and will enable the next generation models to be smarter and more practically replicate real-flight vehicles.

Some of the significant improvements in the NGEM for the purpose of improving the aerothermal capabilities of the plasma preheating technology to reliably replicate planetary reentry surface temperatures with high degree of confidence are presented in **Table 1**. This novel invention is able to accurately replicate the planetary reentry surface temperatures and any associated hypersonic flow characteristics within the boundary layer.

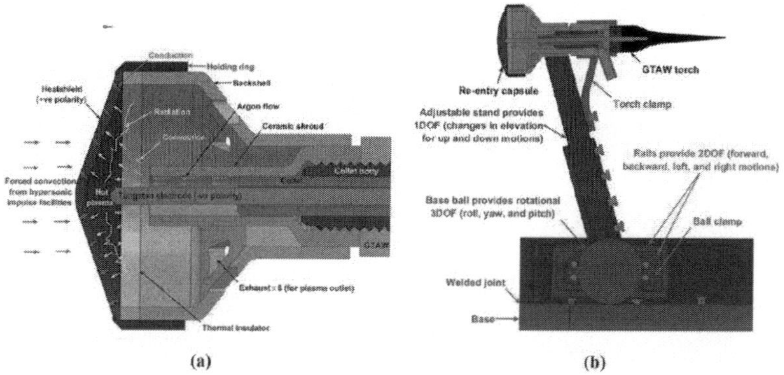

Figure 19.
Sectional view showing improved operational capabilities for the NGEM [36]. (a) Thermophysics and heat transfer. (b) Full view of the NGEM.

Parts	Present work	Future work	Scientific relevance for future work
Backshell thickness	5.0 mm	2.5 mm	Reduced mass and heat-sink effects
Geometry	Conventional	Advanced	Better thermodynamic performance
Thermal barrier system	Absent	Present	Reduced conduction losses
Holding ring	Mild steel	TiC	Higher thermal capability over steel
Backshell	Mild steel	Mild steel	Housing and support
Ceramic shroud	96% Alumina	Zirconia	Higher thermal capability over 96% Alumina
GTAW torch	Separated	Attached	Thermal optimisation over detached features
Stand	Fixed	Adjustable	Flexibility in elevations
6DOF	Absent	Present	Better aerodynamic manoeuvrability

Table 1.
Materials selection for components and parts for hypersonic/ablation experiments [36].

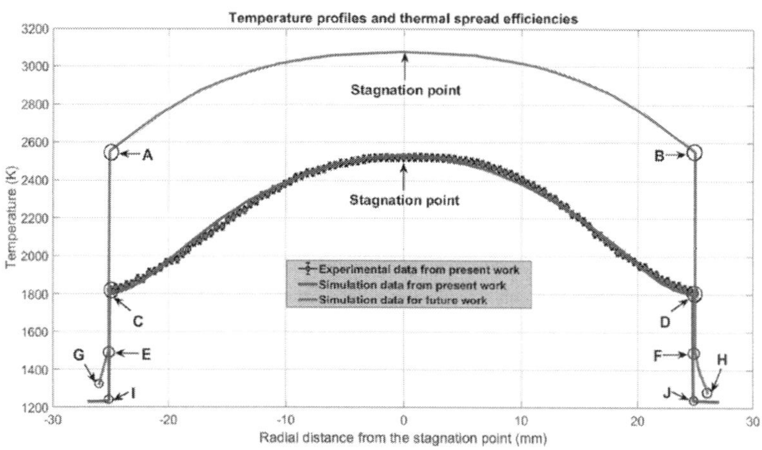

Figure 20.
Spatial temperature profiles showing heatshield thermal spread [36].

Figure 20 shows the temperature profile from FEA simulation using Ansys Workbench. The contact regions in the future work have been designed to minimise conduction losses at heatshield edges (shoulder regions) using a thermal barrier material (Zirconia). The improvement of surface temperature profile for future work can be seen in **Figure 20**. The details from spatial temperature profiles for the present work have been extensively published by the author [27].

6. Conclusions

The major area of interest is using plasma to preheat probe models that can be used for ablation studies in cold flow hypersonic wind tunnels. This novel plasma preheating technology is capable of generating the needed heat flux for the surface temperature characterisation without using any Arc-Jet or plasmatron facilities. This technique aims to produce a newer and better method of aerothermodynamic tests for investigating ablation samples of reentry probes in expansion tubes. The probe used in the present work was very similar to the European Standard Probe with a stagnation temperature of about 2500 K. The advantages that are associated with this newly innovated plasma preheating technique include: (1) light weight and portability of model; (2) surface temperature control; (3) the ability to replicate entries for different planetary missions due to its capability to perform well in most types of reentry gases like O_2 N_2, Air, CO_2, He, Ne, Ar, H_2, CH_4, NH_3, etc.; (4) ability to be applied to different types of heatshield materials like PICA, SIRCA, Avecoats, C/C Composites, graphite, etc.; (5) ability to be used for different axi-symmetric payload geometries like Stardust, Orion, Hayabusa, SpaceX-dragon, etc.; (6) ability to perform well in both short and long duration wind tunnels including shock tunnels; and (7) highly economical and less than 10% of the operational costs of running the high enthalpy plasmatron and NASA Ames Arc-Jet facilities. This work presents reasonable contributions to reentry studies under conditions that replicate characteristics of hypersonic reentry flights. Future applications for the technique are expected to be found in hypersonic impulse facilities that can simulate the true flow energy under reentry conditions.

References

[1] Hippke, M., 2019. Spaceflight from Super-Earths is difficult. International Journal of astrobiology, *18*(5), pp.393-395.

[2] Fusaro, R., Ferretto, D. and Bader, P., 2019. Preliminary design of a Thermal and Energy Management Subsystem for hypersonic vehicle concepts.

[3] Boehrk, H., Weihs, H. and Elsäßer, H., 2019. Hot Structure Flight Data of a Faceted Atmospheric Reentry Thermal Protection System. *International Journal of Aerospace Engineering, 2019*.

[4] Julian, H. and Eggers, A.J., 1958. A study of the motion and aerodynamic heating of ballistic missiles entering the earth's atmosphere at high supersonic speeds. NACA Rept, *1381*.

[5] Loehle, S., Staebler, T., Reimer, T. and Cefalu, A., 2015. Photogrammetric surface analysis of ablation processes in high-enthalpy air plasma flow. AIAA Journal, *53*(11), pp.3187-3195.

[6] Loehle, S., Fasoulas, S., Herdrich, G. H., Hermann, T.A., Massuti-Ballester, B., Meindl, A., Pagan, A.S. and Zander, F., 2016. The plasma wind tunnels at the institute of space systems: current status and challenges. In *32nd AIAA Aerodynamic Measurement Technology and Ground Testing Conference* (p. 3201).

[7] Leyland, P., McIntyre, T., Morgan, R., Jacobs, P., Zander, F., Sheikh, U., Eichmann, T., Fahy, E., Joshi, O., Duffa, G. and Potter, D., 2013. Radiation-ablation coupling for capsule reentry heating via simulation and expansion tube investigations. In *Proceedings of the 5th European Conference for Aeronautics and Space Sciences (EUCASS)* (No. CONF).

[8] Lu, F.K., Wilson, D.R., Bakos, R.J. and Erdos, J.I., 2000. Recent advances in detonation techniques for high-enthalpy facilities. AIAA journal, *38*(9), pp.1676-1684.

[9] Wang, Z.H., Yu, Y.L. and Bao, L., 2018. Heat transfer in nonequilibrium flows with homogeneous and heterogeneous recombination reactions. AIAA Journal, *56*(9), pp.3593-3599.

[10] Gülhan, A., Esser, B., Koch, U., Fischer, M., Magens, E. and Hannemann, V., 2018. Characterization of High-Enthalpy-Flow Environment for Ablation Material Tests Using Advanced Diagnostics. AIAA Journal, *56* (3), pp.1072-1084.

[11] Mortensen, C.H. and Zhong, X., 2016. Real-gas and surface-ablation effects on hypersonic boundary-layer instability over a blunt cone. AIAA Journal, *54*(3), pp.980-998.

[12] Poovathingal, S., Schwartzentruber, T.E., Murray, V.J. and Minton, T.K., 2016. Molecular simulation of carbon ablation using beam experiments and resolved microstructure. AIAA Journal, *54*(3), pp.999-1010.

[13] Lei, L., Guangyue, D., Lei, Z., Zhenfeng, W. and Yewei, G., 2019. Experimental Model Design and Preliminary Numerical Verification of Fluid–Thermal–Structural Coupling Problem. AIAA Journal, *57*(4), pp.1715-1724.

[14] Sagnier, P. and Verant, J.L., 1998. Flow characterization in the ONERA F4 high-enthalpy wind tunnel. AIAA journal, *36*(4), pp.522-531.

[15] Sheikh, U.A., Morgan, R.G. and McIntyre, T.J., 2015. Vacuum ultraviolet spectral measurements for superorbital earth entry in X2 expansion tube. Aiaa Journal, *53*(12), pp.3589-3602.

[16] Horvath, T.J., Berry, S.A. and Merski, N.R., 2004. *Hypersonic*

boundary/shear layer transition for blunt to slender configurations-A NASA Langley experimental perspective. National Aeronautics and Space Administration Hampton Va Langley Research Centre.

[17] Asma, C.O., Tirtey, S. and Schloegel, F., 2012. Flow Topology Around Gas, Liquid and Three-Dimensional Obstacles in Hypersonic Flow. AIAA journal, *50*(1), pp.100-108.

[18] Eswein, N., Herdrich, G., Fasoulas, S. and Röser, H.P., 2011. Investigation of Graphite Ablation at IRS. In *42nd AIAA Thermophysics Conference* (p. 3615).

[19] Ostrom, C., Greene, B., Smith, A., Toledo-Burdett, R., Matney, M., Opiela, J., Marichalar, J., Bacon, J. and Sanchez, C., 2019. Operational and Technical Updates to the Object Reentry Survival Analysis Tool.

[20] Scala, S. M. (1962), The Ablation of Graphite in Dissociated Air: *Theory*, Missile and Space Division, General Electric, pp. 1-94.

[21] Banerji, N., Leyland, P., Fahy, E. and Morgan, R., 2018. Venus Entry Flow over a Decomposing Aeroshell in X2 Expansion Tube. Journal of Thermophysics and Heat Transfer, *32* (2), pp.292-302.

[22] Zander, F., Morgan, R.G., Sheikh, U., Buttsworth, D.R. and Teakle, P.R., 2013. Hot-wall reentry testing in hypersonic impulse facilities. AIAA journal, *51*(2), pp.476-484.

[23] Lewis, S.W., Morgan, R.G., McIntyre, T.J., Alba, C.R. and Greendyke, R.B., 2016. Expansion tunnel experiments of earth reentry flow with surface ablation. Journal of Spacecraft and Rockets, *53*(5), pp.887-899.

[24] Alba, C.R., Greendyke, R.B., Lewis, S.W., Morgan, R.G. and McIntyre, T.J., 2016. Numerical modeling of earth reentry flow with surface ablation. Journal of Spacecraft and Rockets, *53* (1), pp.84-97.

[25] Glass, D. (2008), Ceramic matrix composite (CMC) thermal protection systems (TPS) and hot structures for hypersonic vehicles, in `15th AIAA International Space Planes and Hypersonic Systems and Technologies Conference', pp. 2682-2718.

[26] Dec, J. and Braun, R. (2006), An approximate ablative thermal protection system sizing tool for entry system design, in `44th AIAA Aerospace Sciences Meeting and Exhibit', pp. 780-794.

[27] Iyinomen Daniel Odion, Ray Malpress, and David Buttsworth, 2021, "Technique Development for Investigating Axisymmetric Ablation Models in Hypersonic Impulse Facilities, " AIAAJ, Vol. 59, No. 6 (2021), pp. 1899-1913 doi: doi/abs/ 10.2514/1.J059629.

[28] Iyinomen, D.O., 2020a, Numerical and experimental analyses of ablation measurements in expansion wind tunnel facilities using a new plasma pre-heating technique. International Journal of Thermofluids, *3*, p.100019.

[29] Iyinomen, D.O., 2020b, Numerical Approach to Ablation Measurements Using A New Plasma Pre-Heating Technique. *International Journal of Thermofluids*, p.100014.

[30] Traidia A. (2011), Multi-physics modelling and numerical simulation of GTA weld pools, PhD thesis, Ecole Polytechnique.

[31] Wang, X., Luo, Y. and Fan, D., 2019. Investigation of heat transfer and fluid flow in high current GTA welding by a unified model. International Journal of Thermal Sciences, *142*, pp.20-29.

[32] Anand, K.R. and Mittal, V., 2017. Review on the parametric optimization of tig welding. International Research

Journal of Engineering and Technology (IRJET), Volume: 04 Issue: 01 January 2017 e-ISSN: 2395 -0056, p-ISSN: 2395-0072.

[33] Siewert, E., Baeva, M. and Uhrlandt, D., 2019. The electric field and voltage of dc tungsten-inert gas arcs and their role in the bidirectional plasma-electrode interaction. *Journal of Physics D: Applied Physics*, *52*(32), p.324006.

[34] Iyinomen, D.O., 2019, Technique development for investigation of axisymmetric graphite sample oxidation in hypersonic flow. *Doctoral dissertation at the Institute for Advanced Engineering and Space Sciences, University of Southern Queensland.*

[35] Kondo, Y., Maruno, H., Tominaga, H., Soya, H. and Etoh, T.G., 2003, August. Ultra-high-speed video camera and its applications. In *25th International Congress on High-speed photography and photonics* (Vol. 4948, pp. 53-58). International Society for Optics and Photonics.

[36] Iyinomen D.O, 2021. Plasma preheating technology for replicating planetary re-entry surface temperatures. The Aeronautical Journal, https://doi.org/10.1017/aer.2021.72

[37] Keenan, J. and Candler, G., 1993, July. Simulation of ablation in earth atmospheric entry. In *28th Thermophysics Conference* (p. 2789).

[38] Sun, X., Yang, H. and Mi, T., 2020. Heat Transfer and Ablation Prediction of Carbon/Carbon Composites in a Hypersonic Environment Using Fluid-Thermal-Ablation Multiphysical Coupling. *International Journal of Aerospace Engineering, 2020.*

[39] Zeng, M., Liu, J., Xu, D. and Zhang, W., 2014. Numerical study of the transient characteristics of ablative hypersonic flow fields. International Conference on Heat Transfer, Fluid Mechanics and Thermodynamics.

Section 3

Vehicle Design Analysis

Chapter 5

Aero Heating Optimization of a Hypersonic Thermochemical Non-Equilibrium Flow around Blunt Body by Application of Opposing Jet and Blunt Spike

Abstract

The goal of this work is to give optimum aerothermal solutions for thermal protection of the nose wall of space shuttles during atmospheric reentry, where the air flow is hypersonic, nonequilibrium reactive flow (vibrational and chemical) behind detached shock waves, it's governed by Navier–Stokes equations with chemical reaction source terms, and modelled using five species (N_2, O_2, NO, N, O) and Zeldovich chemical scheme with five reactions. This study which simulates the flow using the software Fluent v.19 focuses on the comparison between three protection techniques based on the repulsion of the shock wave, the first is geometric, it consists in introducing a spike that makes the right shock move away from the nose of the shuttle, this allows the endothermic physicochemical processes of dissociation and ionization to absorb heat, the second technique is based on an opposite jet configuration in the frontal region of the nose, this jet allows to push the strong shock, and consequently reduce the heat released, the last technique is the assembly of the two previous techniques; Jets nearby the spike noses were set up in front of the blunt body to reconfigure the flow field and reduce aerodynamic overheating. The opposing jet model reduces the heat at the nose by 12.08% compared to the spike model and by 20.36% compared to the spike jet model. The flow field reconfiguration was the most important factor in heat reduction, according to the quantitative analysis, a combination parameter was given as the main criterion for designing spiked bodies with opposing jets for the goal of heat reduction based on the locations of the reattached shock and its interaction with the conical shock. The results obtained are in good agreement with the specialized literature.

Keywords: hypersonic reactive flows, non-equilibrium, dissociation, opposing jet, blunt spike, Lobb sphere

1. Introduction

The concept of hypersonic flight has attracted worldwide attention since it was proposed in the 1940s [1–3]. Since there, different hypersonic vehicles were developed as well as hypersonic missiles, aircraft and re-entry vehicles. However, the significant number of technical challenges have surfaced which are critical to the successful development of these high-speed flight vehicles. Understanding, analyzing and predicting high-speed flow around blunt bodies pose a practical and important engineering problem; faster and better design of new flight vehicles depends on it. The high heat of a hypersonic aircraft during flight imposes severe demands on the materials and structures, so the reduction of heat transfer rate plays an important role during the conceptual design of re-entry vehicles. Classical thermal protection systems such as ablatives [4] are less adapted to the rapid growth of spacecraft technology: ablatives are related to the coating thickness, and are not convenient for shape change. To improve the flow field in front of a vehicle nose, additional solutions such as active cooling approach aero-spikes [5] and opposing jets [6] have been developed. Different strategies exhibit varying characteristics in a hypersonic flow field.

The concept of a spiked blunt body was first proposed by Bogdonoff [7]. Flow separation in front of blunt bodies at supersonic speeds, have been made since the early 1950s [8–11]. Spikes have been shown to create a separation zone over blunt bodies, lowering the aerodynamic heat rate and pressure distribution, which is beneficial for thermal protection and drag reduction. In the 1960s, separation characteristics and the resulting flow instability have been the subject of extensive research. Maull explored the effects of spike length and shape on the flow field properties of blunt bodies, concluding that flow oscillation was generated by two factors: shock wave-induced separation and flow reattachment [12]. Using a mix of spike lengths and cone angles, wood explored spiked cone cylinders flying at Mach 10 and established five possible flow patterns with related scopes [13]. Reding et al. examined unstable aerodynamics for a spiky drag reduction device based on structure deflection coupled with thermal expansion generated by aerodynamic heating. Many studies on numerical simulations and solutions have allowed for a great level of insight with regards to spiked blunt bodies since the rapid rise of computer technology began in the 1990s. Many of these researches were validated by collaboration between experiments and numerical simulations. Mehta investigated the relationship between the aerodynamic heat flux and pressure distribution over spiked bodies at Mach 6.8 [14].

The opposing jet technique was introduced initially by Lophtoff [15] and Warren [16], and its obvious effects on drag and heat reduction for aircraft were realized. The interaction between the opposing jet and the free stream determines the flow pattern. Hayashi et al. [17, 18] investigated the opposing jet using both experimental and numerical methods in several investigations. Their tests were carried out in a traditional blow-down type wind tunnel with a free stream of Mach 3.98, and the axisymmetric Navier-Stokes equations were solved using the implicit finite difference method. Their research revealed that the ratio of the opposing jet's stagnation pressure to that of the free stream had a significant impact on the flow mode. Isao Tamada et al. [19] investigated the heating reduction of the ogive body and hemispherical nose cylinder body experimentally and numerically, at M = 3.98 and M = 8.0. They found that local Reynolds number management and recompressed shock monument are critical to reducing aerodynamic heating, and that the ogive body was more effective at reducing heating with the same mass flow rate because of its large enough recirculation region to cover the entire nose tip.

Huang et al. [20, 21] investigated some opposing jet configurations with other cooling approaches, such as spike, aerodisk, and forward-facing cavity, and discovered combined promising drag and heat flux reduction effects, as well as the coupling mechanism between the self-sustained oscillations induced by the jet and the unsteady modes induced by the other configurations. An experiment was conducted by Jiang et al. [22] which a new concept of the non-ablative thermal protection system for hypersonic vehicles was first proposed, and the blunt spike was combined with lateral jets for developing a shock reconstruction system at the front side of hypersonic vehicles, to achieve effective wave for reducing drag under non-zero attack angles and also to avoid severe aero heating (rocket).

In this work, a blunt re-entry vehicle is modeled and analyzed in ANSYS Fluent 19 which represents the distribution of heat flux on the surface of a representative Lobb sphere blunt body and the coupled effects of thermochemical non-equilibrium and chemical reactions on the hypersonic air flows. Validation is performed with the obtained CFD results which are in good agreement with the experimental values of Liu and Jiang [23] for blunt spike body and Hayashi et al. [24] for opposing jet. Next, optimization is carried out by placing a jet at the front of the blunt spike body. The main simulation results are discussed by comparing the spike and jet heat reduction configurations.

2. Mathematical and physical modeling

The two-dimensional, steady, compressible, Navier Stokes equation set was applied as the governing equations. The fundamental governing equations are represented in Eqs. (1)–(6). The density was calculated using the ideal gas law. For viscous and compressible flow, the viscosity is generally [25] calculated as a function of temperature as defined by Armally and Sutton law as in Eq. (7), and hence it is used in the present work.

2.1 Species conservation

For species in a mixture, the mass conservation equation is dictated by:

$$\underbrace{\frac{\partial \rho_\alpha}{\partial t}}_{1} + \underbrace{\frac{\partial}{\partial x}\left(\rho_\alpha u - \left(-q^D_{\alpha x}\right)\right)}_{2} + \underbrace{\frac{\partial}{\partial y}\left(\rho_\alpha u - \left(-q^D_{\alpha y}\right)\right)}_{3} = \underbrace{\dot{w}_\alpha}_{4} \tag{1}$$

The global continuity equation is given by

$$\underbrace{\frac{\partial \rho}{\partial t}}_{1} + \underbrace{\frac{\partial}{\partial x}(\rho u)}_{2} + \underbrace{\frac{\partial}{\partial y}(\rho u)}_{3} = 0 \tag{2}$$

2.2 Global momentum conservation

The mixture's momentum balance is given by

$$\underbrace{\frac{\partial \rho u}{\partial t}}_{1} + \underbrace{\frac{\partial}{\partial x}\left(\rho u^2 + p - \tau_{xx}\right)}_{2} + \underbrace{\frac{\partial}{\partial y}\left(\rho u v - \tau_{xy}\right)}_{3} = 0 \tag{3}$$

$$\underbrace{\frac{\partial \rho v}{\partial t}}_{1}+\underbrace{\frac{\partial}{\partial x}\left(\rho u v-\tau_{xy}\right)}_{2}+\underbrace{\frac{\partial}{\partial y}\left(\rho v^{2}+p-\tau_{yy}\right)}_{3}=0 \tag{4}$$

2.3 Vibrational energy conservation

Vibrational energy conservation is a phenomenological characterization of the average energy in each molecular species' vibrational mode. The conservation of vibrational energy is regulated by:

$$\underbrace{\frac{\partial}{\partial t}\left(\rho_m e_m^{vib}\right)}_{1}+\underbrace{\frac{\partial}{\partial x}\left(\rho_m e_m^{vib} u-q_{mx}^{vib}\right)}_{2}+\underbrace{\frac{\partial}{\partial y}\left(\rho_m e_m^{vib} v-q_{my}^{vib}\right)}_{3}=\underbrace{\Omega_m}_{4} \tag{5}$$

2.4 Total energy conservation

Total energy conservation (internal + kinetic) is regulated by

$$\underbrace{\frac{\partial(\rho E)}{\partial t}}_{1}+\underbrace{\frac{\partial}{\partial x}(\rho E+p)u-\left(u\tau_{xx}+v\tau_{\gamma x}+qt_{x}\right)}_{2}+\underbrace{\frac{\partial}{\partial y}\left((\rho E+p)v-\left(u\tau_{xx}+v\tau_{\gamma x}+q_{\gamma}t\right)\right)}_{3} \tag{6}$$

2.5 Armaly Sutton law

$$\mu_{\alpha}=0.1\ exp\left[\left(A_{\alpha}^{\mu}\ ln\ (T)+B_{\alpha}^{\mu}\right)ln\ (T)+C_{\alpha}^{\mu}\right] \tag{7}$$

2.6 Chemical kinetic model

A thermo-chemically non-equilibrium flow of a five-component air model consisting of species N_2, O_2, NO, N and O was considered. We have made the hypothesis of a chemical flow at vibrational equilibrium; the ionization phenomena have been neglected. The most important chemical reactions between these species are: [26, 27].

$$\mathrm{N_2+M \rightleftharpoons 2\,N+M} \tag{8}$$

$$\mathrm{O_2+M \rightleftharpoons 2O+M} \tag{9}$$

$$\mathrm{NO+M \rightleftharpoons N+O+M} \tag{10}$$

$$\mathrm{NO+O \rightleftharpoons N+O_2} \tag{11}$$

$$\mathrm{N_2+O \rightleftharpoons NO+N} \tag{12}$$

The chemical source names are formed from reactions that take place between the gas's constituents. A mass transfer mechanism occurs between species as reactions occur so the formulas for these mass transfer rates are determined. Several separate elementary chemical reactions between species in the gas can take place at the same time. Consider the rth chemical reaction of N_r elementary reactions between N_s chemically reacting species:

$$\sum v'_{\alpha,r}X_{\alpha}=\sum v''_{\alpha,r}X_{\alpha} \tag{13}$$

There is a forward and backward portion to the chemical reaction equation, Eq. (13). The forward and backward reaction rates are calculated as follows:

Forward:

$$\frac{d[X_\alpha]_r^f}{dt} = \left(v''_{\alpha,r} - v'_{\alpha,r}\right)\left[k_{f,r}\prod_{\alpha=1}^{N_s}[X_\alpha]^{v'_{\alpha,r}}\right] \tag{14}$$

Backward:

$$\frac{d[X_\alpha]_r^b}{dt} = \left(v''_{\alpha,r} - v'_{\alpha,r}\right)\left[k_{b,r}\prod_{\alpha=1}^{N_s}[X_\alpha]^{v''_{\alpha,r}}\right] \tag{15}$$

Where

- $k_{f,\,r}$ and $k_{b,\,r}$ are the forward and backward reaction rate coefficients of reaction r, which are both affected by the temperature of the reaction.

The net rate for the above general reaction r can be written as

$$\frac{d[X_\alpha]}{dt} = \frac{d[X_\alpha]_r^f}{dt} - \frac{d[X_\alpha]_r^b}{dt} = \left(v''_{\alpha,r} - v'_{\alpha,r}\right)\left[k_{f,r}\prod_{\alpha=1}^{N_s}[X_\alpha]^{v'_{\alpha,r}} - k_{b,r}\prod_{\alpha=1}^{N_s}[X_\alpha]^{v''_{\alpha,r}}\right] \tag{16}$$

The equation above is a general form of the law of mass action, which assures that total mass is preserved during a chemical reaction.

The FLUENT uses the expression given by the law of ARRHENIUS to calculate the direct speed constant. The expression of ARRHENIUS is given by:

$$K_{f,r} = C_{f,r} T_{f,r}^{n_{f,r}} e^{\frac{-E_{f,r}}{kT_{f,r}}} \tag{17}$$

3. Numerical modeling

For aerodynamic flow, Reynolds-Average Navier-Stokes (RANS) equations are adopted as the governing equations. The convective terms are approximated by the AUSM-DV scheme [28] with a MUSCL approach to increase the numerical accuracy. The turbulence model is required for several shear layers including the boundary layer, The k-ω model (SST) [29] model in the region close to the wall was used.

3.1 Computational model

The structures of a Lobb sphere blunt body are shown in **Figure 1**. The diameter of the body is 6.35 mm with a length of 1.3 mm [30]. The geometry in **Figure 1** was created in a way that the simulation will be run using the axisymmetric Navier-Stock's equations, therefore, a two-dimensional symmetric geometry is created, with an axis defined at the radial centre of the studied body. This technique makes it possible to reduce the computational domain used and consequently reduce the calculation time.

Then we set up an opposing jet in the same Lobb sphere with a radius of 0.5 mm as shown in **Figure 2**.

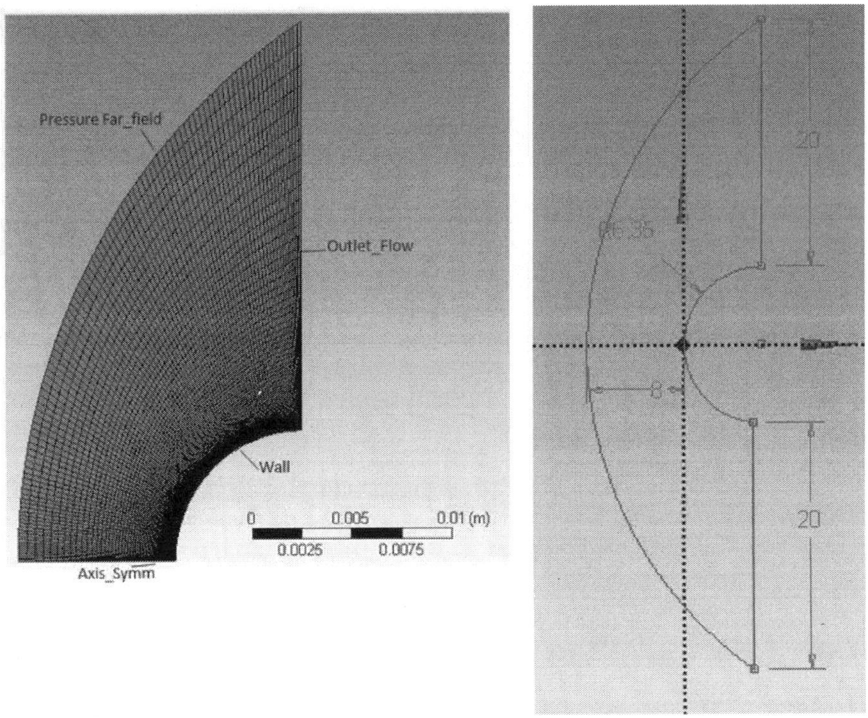

Figure 1.
Lobb sphere mesh and geometry.

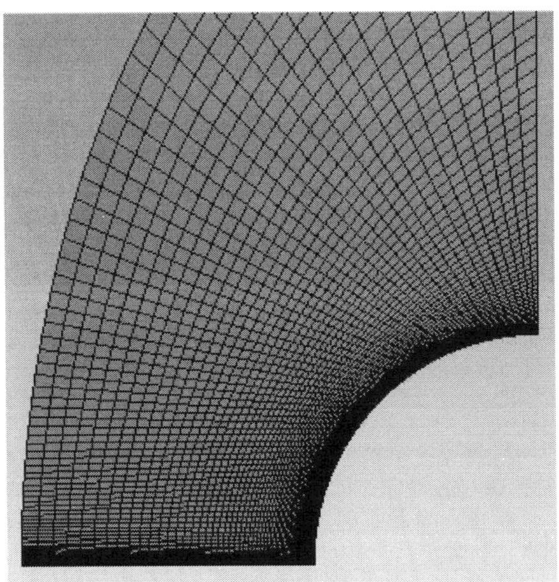

Figure 2.
Lobb sphere with opposing jet mesh and geometry.

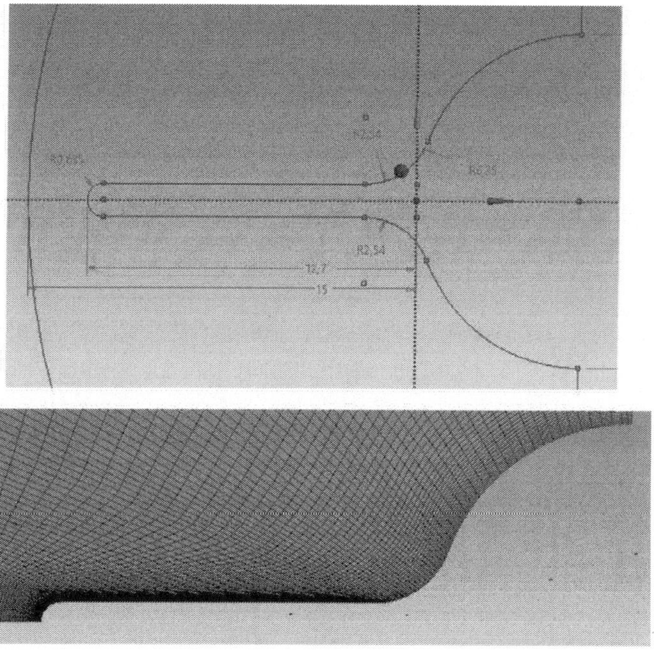

Figure 3.
Blunt spike mesh and geometry.

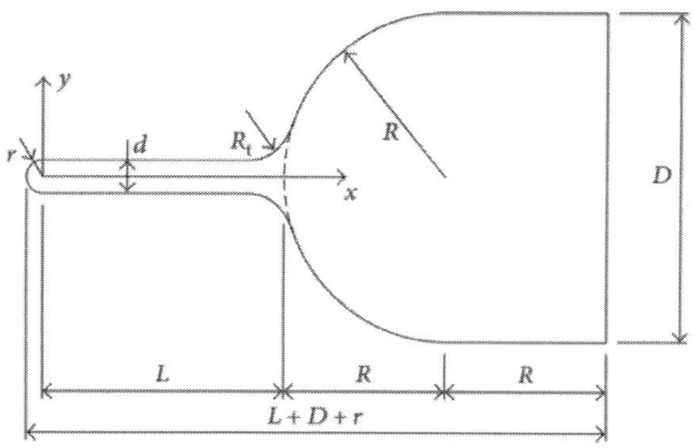

Figure 4.
Configuration of the spiked blunt body for simulation [31].

Next, we install the second heat reduction configuration, the blunt Spike in front of the same Lobb sphere body profile to reconfigure the flow field and reduce the overheating in a re-entry hypersonic flight as shown in **Figure 3**. **Figure 4** shows the configuration and dimensions of the spiked (4) [31].

- The characteristic dimensional are d/D = 0.1, L/D = 0.9 and Rt/D = 0.2 where:

  D: is the diameter of the spherical head profile, d: is the diameter of the spike,

  L: is the spike length and Rt: is the transition part radius at the spike root.

3.2 Mesh independence

An appropriate grid appears to be the key to numerical prediction accuracy, particularly in the case of aerodynamic heating prediction. A grid independence investigation was undertaken over multiple grid densities before initiating the CFD simulations. Because the heat transfer between the surrounding air and the vehicle wall was so important in this research, attention has been devoted to the near-wall mesh quality.

In the Lobb sphere blunt body, five-level grids are employed, and the details of these grids are shown in **Table 1**. The region located near the wall has meshed with a gradient structured mesh. The flow temperature for different grids is shown in **Figure 5**, we notice that the more the grid is refined more the temperature is stable where the result starts to be independent of the mesh. As computational time depends on the size of the grid. Thus the grid of case 4 is used in the following simulations with considering the calculation efficiency.

Adaption	Initial	Adaption1	Adaption2	Adaption3	Adaption4	Adaption5
Cells	2500	4858	10,570	25,146	80,974	129,963
Nodes	5601	7054	15,990	11,575	37,546	89,125
T(K)	8743.415	9193.686	9691.146	9808.36	10,130	10,280

Table 1.
Lobb sphere grid independency.

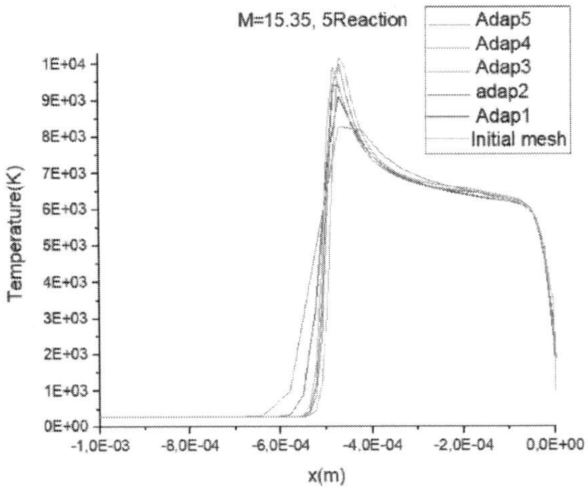

Figure 5.
Lobb sphere grid independency.

Adaption	Initial	Adaption1	Adaption2	Adaption3
Cells	5000	14,330	52,787	53,360
Nodes	14,626	14,626	53,376	53,948

Table 2.
Blunt spike configuration grid independency.

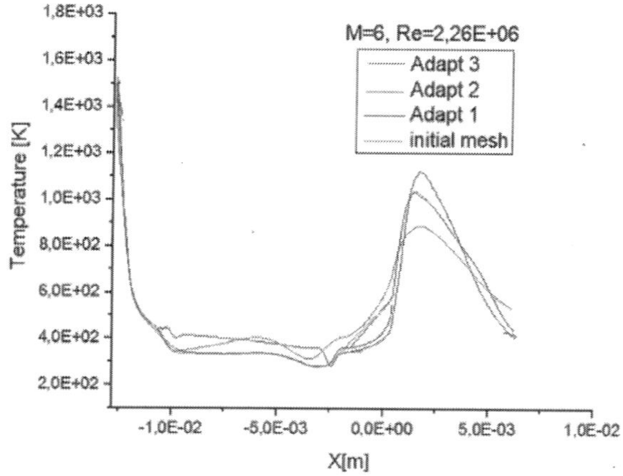

Figure 6.
Blunt spike grid independency.

Always by the same refinement method, the sensitivity test for the spike mesh was performed for three different mesh densities (**Table 2**). We notice that the result become independent of mesh from adapt 2 because the variation of temperature is no longer observed during the refinement in **Figure 6**, from adapt 2 with a temperature of 11123 k to adapt 3 with temperature 11232 k so we can consider that the "adapted 2 mesh" is the optimum mesh.

4. Validation of numerical models

4.1 Validation of Lobb sphere blunt body

By comparing our result of **Figure 7** with that of Tristan [30] which shows the variation of the temperature along with the relaxation range, one notices a good agreement between the two results in terms of pace and the quantitative terms. The slight difference is because Tristan considered a flow out of vibrational equilibrium. The study was simulated under the flowing free stream conditions (**Table 3**).

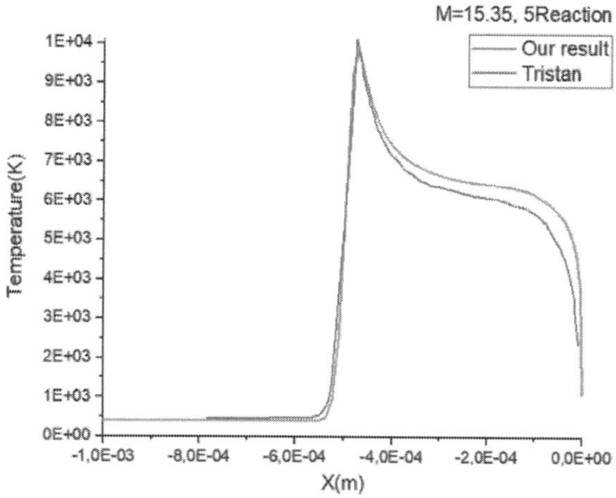

Figure 7.
Temperature along the stagnation ling.

M_∞	Reaction.N	$T_\infty[k]$	$\rho_\infty(\text{kg/m}^3)$	P_∞ [Pa]	K	ω
15.35	5	293	$7.896\ 10^{-3}$	664	3245.962	64.607

Table 3.
Lobb sphere free stream boundary condition [30].

Source	Shock position [mm]
Present results	0.536
Tristan	0.535
Tchuen	0.531
Joly et al	0.598
S'eror	0.557
Lobb	0.552 ± 0.032

Table 4.
Position of the shock of the Lobb sphere [30].

Table 4 illustrates that this result agrees well with the experimentally determined value with an error of 2.89% [32] and the numerical values obtained in the literature [33, 34].

4.2 Validation of spike blunt configuration

The numerical solution validation was carried out by comparing the experimental flow field as shown in **Figure 8**. The conditions were set as $M_\infty = 6$, Re = 2.26E+2 and L/D = 1.0 [31]. Numerical and experimental results are in agreement, indicating the main flow field structures, including conical shock, reattached shock, slip line, shear layer, and separation zone.

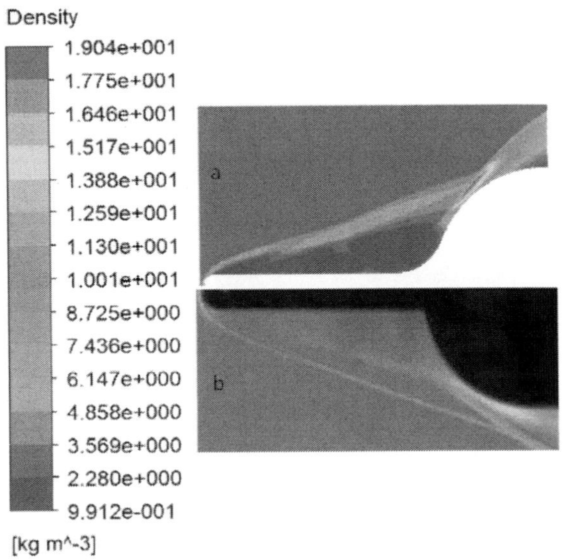

Figure 8.
Flow field wave structure of the spiked blunt body with spike length L/D = 1.0; (a) our numerical photograph result; (b) experimental Schlieren photograph [23].

4.3 Validation of opposing jet configuration

Numerical results are compared with the experiment by NASA Langley in 1987 [24]. The radius of the cylinder is 25.4 mm and the thickness of the steel is 12.7 mm [35] (**Figure 9**). The free flow parameters are as follows (**Table 5**).

To investigate the effect of thermal protection systems we carried out a comparison between the opposing jet model and spike model at 26 km under the following free stream conditions illustrated in **Table 6**.

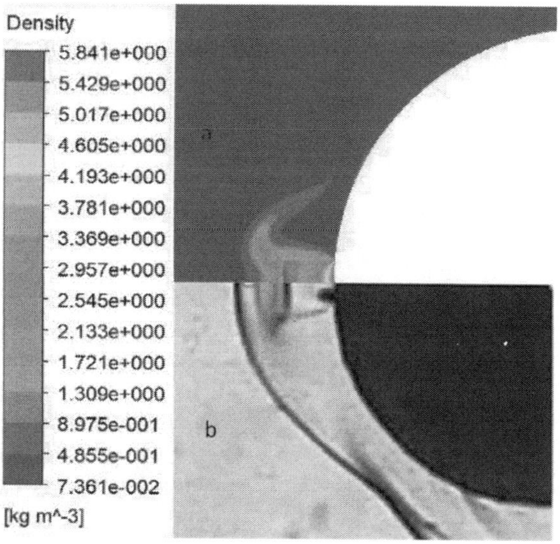

Figure 9.
Flow field wave structure of opposing jet; (a) our numerical photograph result; (b) experimental Hayashi photograph [24].

Free stream	The opposing jet	The wall
Gas: Air Mach number: 3.98 Total pressure: 1.37 MPa Total temperature: 397 K	Gas specified in mole fraction: 0.2571 CO_2, 0.3142 N_2, 0.4287 $H_2O(g)$ Mach number: 1 Total pressure ratio: 0.4 Total temperature: 200 K	Temperature 295 K

Table 5.
Opposing jet free stream boundary condition [35].

Free stream	Opposing jet		Wall condition
• Gas: air • Mach number: 6 • Total pressure: 4.02 MPa • Total temperature: 1812 K	Pressure ratio: 1.25 Total temperature: 300 K	**Sizing** 1. Opossing jet diameter 1 mm 2. Assembly of spike and jet 0.15 mm	Convection with heat transfer coefficient 1577 (W/m^2 K) according to [36, 37]

Table 6.
Boundary condition and sizing.

5. Discussion and results

5.1 Spike effect

We observed that the shock wave scattered along the length and decreases the temperature in the vehicle nose surface from 1254 to a value of 1053 k from spike nose to the body nose respectively as noticed in **Figure 10**. The presence of the spike, blunt transforms the bow shock into a weaker conical shock.

Figure 10.
Blunt spike temperature distribution.

Figure 11.
Blunt spike Mach number distribution.

Figure 12.
Blunt spike pressure distribution.

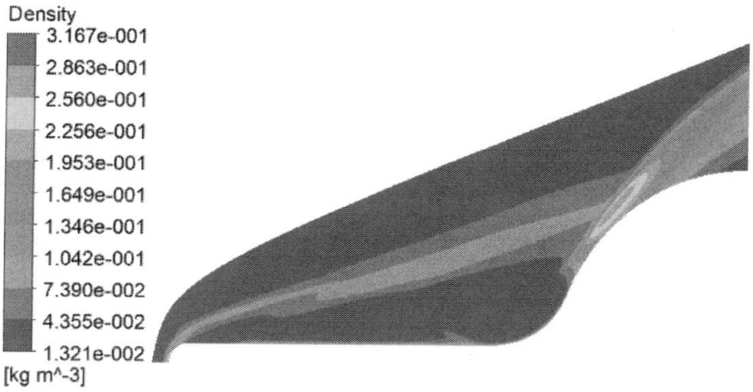

Figure 13.
Blunt spike density distribution.

The high-speed flow traveled to the blunt body shoulder as shown in **Figure 11**, result in a reattached shock formation with a high-pressure zone in **Figure 12**, the unfavorable pressure gradient inverses the gas flow to the spike nose. Thus, a circumfluence zone was generated around the spick and the blunt body nose which induce a lower velocity than that after the conical shock, generating a shear layer.

represented in **Figure 13**, which reduces some of the fluid in the circumfluence area. So, the combined effect of the reduced foreshock and the recirculation zone on the main body can result in significant reductions in aero heating estimated by 16.67%.

5.2 Jet effect

Figure 14 represent the distribution of pressure, temperature, density and streamlines of the fluid field with opposing jet configuration shows that the jet layer reattaches to the blunt body surface, and then a low temperature recirculation region is formed lead to the formation of a normal shock ward off from the blunt body nose. Where a recompression shock wave also formed downstream the reattachment region of the jet layer.

The interaction between the flow field and the solid structure determines the heating rate distributions. The temperature difference between the solid external surface and the fluid layer near the fluid-solid interface will change as the temperature field of the solid structure changes due to aerodynamic heating, which will affect in turn the heat transfer between the fluid and the solid domain. so, by introducing the opposing jet configuration the temperature was reduced from 1035 without jet to 854 with jet model, this led to a heat flux reduction estimated by 28.75%.

5.3 Comparison between spike and jet configuration

Figure 15 shows that spike configuration ward off the shock wave from the blunt body nose by 0.01 m were the spike nose faces the maximum temperature by reducing the blunt body temperature by 16.67% from the blunt nose body, this configuration, by its virtue of geometry and nose sizing which are smaller compared to blunt body nose, admit the use of refractory material that can withstands very high temperatures such as zirconium only on the nose of the spike instead of all the blunt body nose, this will reduce the economic cost of the spacecraft design. Compared with jet cooling configuration, the direct interaction between the jet gaze's

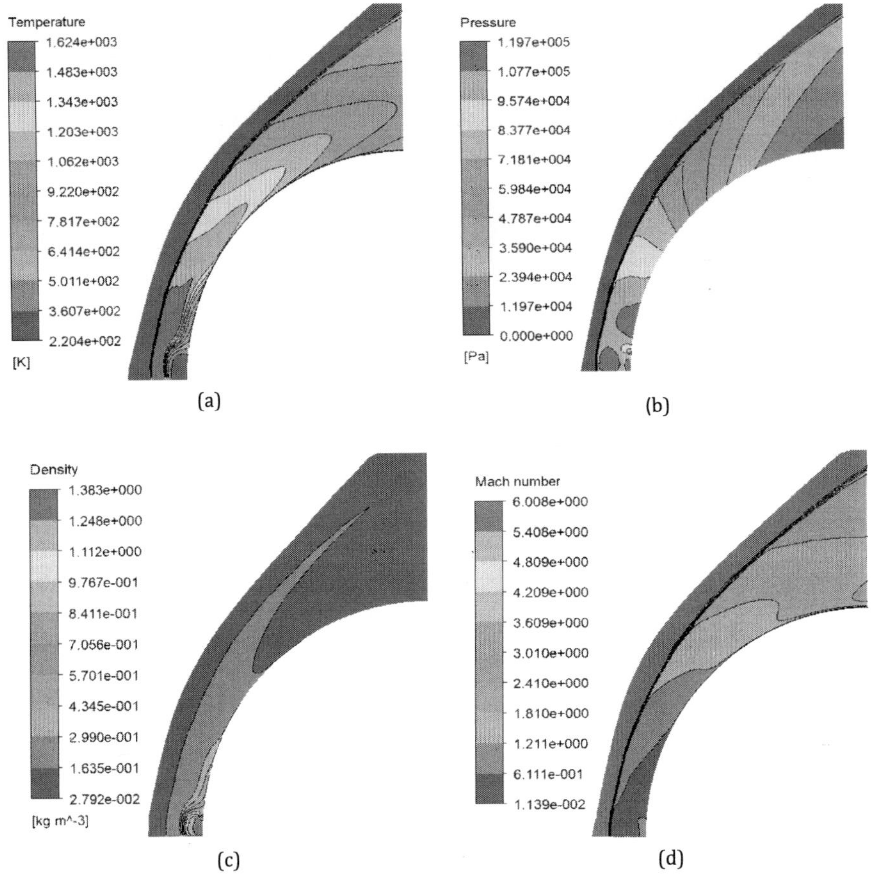

Figure 14.
Jet configuration properties. (a) Opposing jet temperature distribution; (b) opposing jet pressure distribution; (c) opposing jet density distribution; (d) opposing jet Mach number distribution.

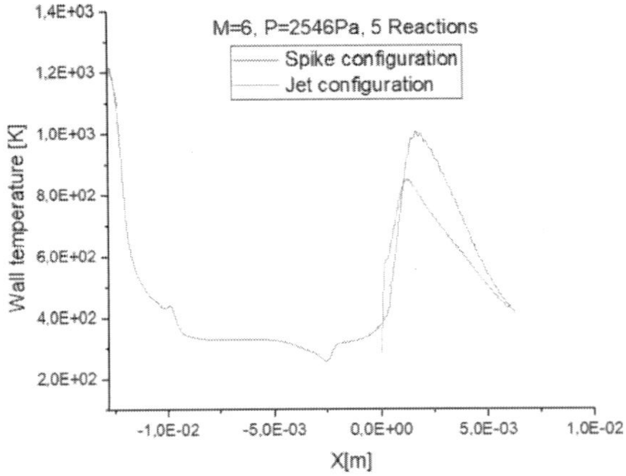

Figure 15.
Opposing jet and bunt spike wall temperature distribution.

and free stream flow, induces a redistribution of the temperature in the mixing zone ward off the normal shock from the nose blunt body about 0.3 mm, which allows the reduction of the wall heat transfer of the body by approximately 28.75%, this cooling system must be associated with important subsystems, particularly in terms of feasibility constrained by the mass and layout problems (regulator, pump, storage tank of gaze, etc.).

5.4 Jet spike configuration

To reduce the huge heat load in the small area of the spike nose, we have proposed an assembly between the spike and the opposing jet configuration by introducing a jet into the nose of the spike to protect the nose from the overheating temperatures and minimize the cost of conventional thermal protection systems such as zirconium. **Figure 16** depicts the velocity of the velocity distribution in the analysis models with and without an opposing jet. The spike generates flow separation in the analytical model without an opposing jet, forming a primary recirculation zone in front of the blunt body. The major recirculation zone in the analytical model with opposing jet is substantially larger than in the model without opposing jet. The reason for this is because the recirculation zones generated by the spike and opposing jet squeeze and interfere with one another, resulting in the formation of a third recirculation zone in the middle. A larger main recirculation zone is formed by these three recirculation zones. This configuration limits the overheat load concentrated just in the frontal nose of the spike as shown **Figure 17**, this limits the cost of the thermal protection system needed in the nose area in the case without jet.

Figure 18 represent the wall temperature for the two configurations spike and the assembly of the two systems, we notice that opposing jet spike have a better effect on the heat flux reduction with 502 K wall temperature compared to the spike configuration with 1205 K in the spike nose area by a reduction estimated by 58.33%.

Instead, the huge reduction in wall temperature is due to the initial jet temperature set of 300 K, where thermal equilibrium occurs between the flow near the wall and the initial jet flow [24].

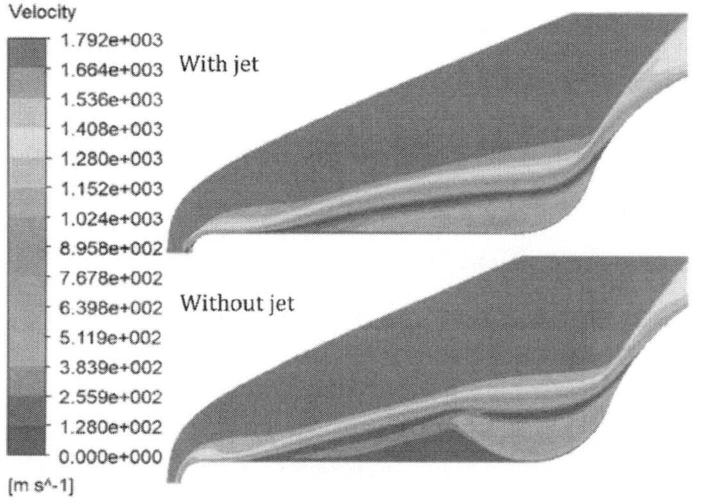

Figure 16.
Velocity distribution of analysis models with and without opposing jet.

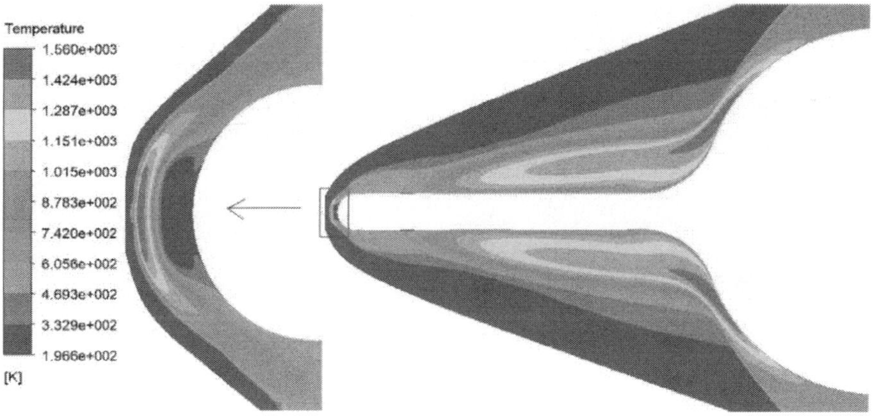

Figure 17.
Jet spike configurations.

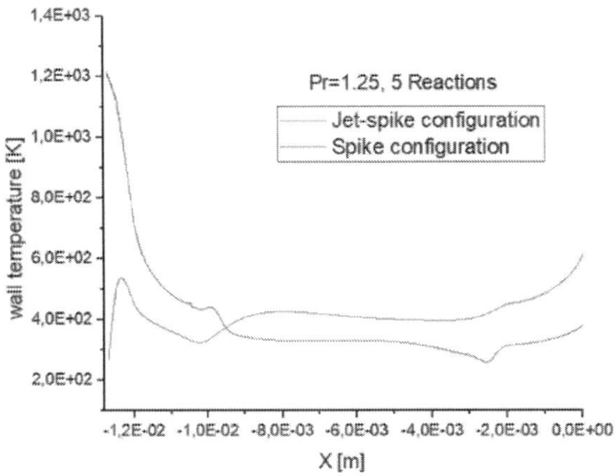

Figure 18.
Wall temperature along with deferent configurations.

6. Conclusion

This study numerically investigates three thermal protection systems that have good potential to reduce the critical heat flux imparted to the blunt-bodies vehicle during its atmospheric reentry, the efficiency of opposing jet, and spike configuration have been discussed and the major conclusions established in this study are summarized as follows:

- The opposing jet configuration isolates the body from the wall from the large hypersonic heat flux, which minimizes the temperature of the stagnation zone in the blunt body of the nose by minimizing the thermal transfer between the flux and the wall by approximately 28.75%, creating an area of recirculation that moves the shock wave slightly away from the tip of the nose; as a result, reducing the heat fluxes transferred to the reentry vehicle walls using opposing jet has a significant interest in the aerospace industry that raises important questions, particularly about the feasibility caused by problems of mass and arrangement layout problem.

- The presence of a spike in a hypersonic flow generates conical shocks in front of the blunt body, a reattached shock settles in the shoulders, and the flow behind the conical shock reattaches to the radial limits of the blunt body, and under the influence of the pressure gradient, it forms a reattached shock. For stable aerodynamic and thermochemical conditions, a phenomenal and endothermic equilibrium is established between the conical shock, the reattached shock and the blunt body, which cause an estimated heat reduction of 16.67% presented in **Figure 15**.

- The jet-spike configuration reduces the heat flux in the spike nose region from 1205 K to 502 K with and without opposing jet respectively, by a reduction estimated by 58.33%.

- In quantitative terms and according to **Figures 15** and **18**, the opposite-jet configuration technique is the best configuration which reduces the near-wall temperature more by 12.08% compared to the spike-configuration, but it remains (spike-configuration) the most practical technique.

Nomenclature

α	chemical species
E	total specific energy (j/kg)
$E_{f,r}$	activation energy for reaction r (J/kg)
e	specific internal energy (J/kg)
$K_{f,r}$	forward reaction rate coefficient in S.I.
$K_{f,b}$	backward reaction rate coefficient in S.I.
N_r	total number of reactions
N_s	total number of species
m	diatomic species
P	pressure (N/m^2)
q	heat flux (J/m^2 s)
q_α^D	mass diffusion flux of species α (kg/(m^2 s))
S	entropy (J/K)
t	time (s)
$T_{f,r}$	temperature of forward reaction r (K)
u, v	speed in x and y directions (m/s)

References

[1] Allouche R, Renane R, Haoui R. Prediction of the optimal speed of an aerospace vehicle by aerothermochemical analysis of hypersonic flow during atmospheric re-entry. Mechanics & Industry. 2020;**21**:208

[2] Allouche R, Haoui R, Renane R. Numerical simulation of reactive flow in non-equilibrium behind a strong shock wave during re-entry into earth's atmosphere. Mechanics & Industry. 2014;**15**:81-87

[3] Anderson JD Jr. Hypersonic and High-Temperature Gas Dynamics. 2nd ed. USA: AIAA Education Series; 2006

[4] Venkatapathy E, Laub B, Hartman GJ, Arnold JO, Wright MJ, Allen GA Jr. Thermal protection system development, testing, and qualification for atmospheric probes and sample return missions: Examples for saturn, titan and stardust-type sample return. Advances in Space Research. 2009; **44**(1):138-150. DOI: 10.1016/j.asr. 2008.12.023

[5] Barzegar Gerdroodbary M, Hosseinalipour SM. Numerical simulation of hypersonic flow over highly blunted cones with spike. Acta Astronautica. 2010;**67**:180-193. DOI: 10.1016/j.actaastro.2010.01.026

[6] Barzegar Gerdroodbary M et al. Transient analysis of counterflowing jet over highly blunt cone in hypersonic flow. Acta Astronautica. 2012;**73**:38-48. DOI: 10.1016/j.actaastro.2011.12.011

[7] Bogdonoff SM. Preliminary investigations of spiked bodies at hypersonic speeds. Journal of the Aerospace Sciences. 1959;**26**(2):65-74

[8] Moeckel WE. Flow separation ahead of blunt axially symmetric body at Mach number 1.76 to 2.10. NACA RM, article E51I25. 1951

[9] Moeckel WE. Flow separation ahead of blunt bodies at supersonic speeds. NACA TN. Vol. 2418. 1951

[10] Stadler JR, Nielsen HV. Heat transfer from a hemisphere-cylinder equipped with flow-separation spikes. NACA TN. Vol. 3287. 1954

[11] Kennedy EC. Calculation of axisymmetric isentropic spike surfaces. Journal of the Aerospace Sciences. 1958; **25**(7):463-464

[12] Maull DJ. Hypersonic flow over axially symmetric spiked bodies. Journal of Fluid Mechanics. 1960;**8**(4):584-592

[13] Wood CJ. Hypersonic flow over spiked cones. Journal of Fluid Mechanics. 1962;**12**(4):614-624

[14] Mehta RC. Numerical heat transfer study over spiked blunt bodies at Mach 6.8. Journal of Spacecraft and Rockets. 2000;**37**(5):700-703

[15] Lopatoff M. Wing-flow study of pressure drag reduction at transonic speed by projecting a jet air from the nose of aprolate spheroid of fineness ratio of 6. Journal of Spacecraft and Rockets. NACA RL51E09. 1951. DOI: 10.1017/S0022112060000694

[16] Warren CHE. An experimental investigation of the effect of ejecting a coolant gas at the nose of a bluff body. Fluid Mechanics. 1960;**8**:400-417. DOI: 10.1017/S0022112060000694

[17] Hayashi K, Aso S, Tani Y. Numerical study of thermal protection system by opposing jet. In: 43rd, AIAA Aerospace Sciences Meeting and Exhibit, AIAA 2005-188. NV: Reno; 2005

[18] Hayashi K, Aso S, Tani Y. Experimental study on thermal protection system by opposing jet in supersonic flow. Spacecraft Rockets.

2006;**43**(1):233-235. DOI: 10.2514/1.15332

[19] Isao T, Shigeru A, Yasuhiro T. Reducing Aerodynamic Heating by the Opposing Jet in Supersonic and Hypersonic Flows, AIAA 2010-991. Florida; 2010

[20] Huang W. A survey of drag and heat reduction in supersonic flows by a counterflowing jet and its combinations. Journal of Zhejiang University Science A. 2015;**16**(7):551-561. Available from: https://link.springer.com/content/pdf/10.1631/jzus.A1500021.pdf

[21] Huang W, Chen Z, Yan L, Yan B, Du Z. Drag and heat flux reduction mechanism induced by the spike and its combinations in supersonic flows: a review. Progress in Aerospace Science. 2019;**105**:31-39. DOI: 10.1016/j.paerosci.2018.12.001

[22] Jiang Z, Liu Y, Han G. A new concept of the non-ablative thermal protection System (NATPS) for hypersonic vehicles. 2013;**51**(3):584-590

[23] Liu Y, Jiang Z. Concept of non-ablative thermal protection system for hypersonic vehicles. AIAA Journal. 2013;**51**(3):584-590

[24] Hayashi K, Aso S, Tani Y. Experimental study on thermal protection system by opposing jet in super-sonic flow. Journal of Spacecraft and Rockets. 2006;**43**(1):233-235. DOI: doi.org/10.2514/1.15332

[25] Armaly B, Sutton K. Viscosity of multicomponent partially ionized gas mixtures. In: 15th Thermophysics Conference. 1980. p. 1495

[26] Dunn MG, Kang S-W. Theoretical and Experimental Studies Reentry Plasmas. Buffalo, N.Y. 14221 Jor Langley Research Center, National Aeronautics and Space Administration° Washington, D.C.: Cornell Aeronautical Laboratory, INC.; 1973

[27] Ma Y, Zhong X. Numerical simulation of transient hypersonic flow with real gas effects. In: 37th Aerospace Sciences Meeting & Exhibit; January 1; University of California, Los Angeles Los Angeles, CA. NV: Reno; 1999. pp. 1-14

[28] Azevedo JLF, Heidi K. 1998 AIAA 1998-2629

[29] Menter FR. Two-equation eddy-viscosity turbulence models for engineering applications. AIAA Journal. 1994;**32**(8):1598-1605

[30] Soubrie T. Prise en compte de l'ionisation et du rayonnement dans la modélisation des′écoulements de rentrée terrestre et martienne. 1er ed. Paris: L'école National Supérieure de L'Aéronautique et de l'espace; 2006

[31] Han G, Jiang Z. Hypersonic Flow Field Reconfiguration and Drag Reduction of Blunt Body with Spikes and Sideward Jets. 4th ed. Beijing, China: State Key Laboratory of High Temperature Gas Dynamics, Institute of Mechanics, Chinese Academy of Sciences; 2017. DOI: 10.1155/2018/7432961

[32] Kennethlobb R. Experimental measurement of shock detachment distance on sphere fired in air at hypervelocity's. In: Nelson WC, editor. Proc. of the AGARD-NATO in The High Temperature Aspect of Hypersonic Flow. 133rd ed. Maryland: Pergamum Press; 1962

[33] Joly V, Coquel F, Marmignon C, Aretz W, Metz S, Wilhelmi H. Numerical Modelling of Heat Transfer and Relaxation in Nonequilibrium Air at Hypersonic Speeds. La Recherche Aerospatiale: Aachen University; 1994

[34] Tchuen G. Modélisation et simulation numérique des écoulements

a haute enthalpie: Influence du déséquilibre électronique. See also AIAA paper 2004–2462 [PhD thesis]. Université d'Aix-Marseille I; 2003

[35] Shen BX, Liu WQ, Yin L. Drag and Heat Reduction Efficiency Research on Opposing Jet in Supersonic Flows. Changsha 410073, China: Science and Technology on Scramjet Laboratory, National University of Defense Technology; 2018

[36] Lester L. Laminar heat transfer over blunt-nosed bodies at hypersonic flight speeds. Jet Propulsion Journal. 1956;**26**: 259-269. DOI: 10.2514/8.6977

[37] Allen HJ, Eggers AJ. A study of the motion and aerodynamic heating of ballistic missiles entering the earth's atmosphere at high supersonic speeds. NACA TR 1381. 1958

Chapter 6

Lifting Entry Analysis for Manned Mars Exploration Missions

Abstract

In the present work, a feasibility study of a manned Mars entry, descent, and landing mission, performed with a lifting vehicle, is analyzed. Mars entry challenges relate to different atmosphere models; consequently, the effective landing capability of a winged configuration is discussed. An entry trajectory study in the Martian atmosphere assuming both a planar and non-planar three degree-of-freedom model is performed. Peak heat rates and time-integrated heat loads during the descent are computed verifying the entry corridor. It is shown that prescribed aerodynamic performances can be modulated explicitly by varying angle of attack and implicitly with bank-angle modulation. Finally, the resulting trajectory is discussed in terms of g-loads, total range performances, and integral heat load absorbed, in the perspective of future manned exploration missions.

Keywords: lifting body, Mars entry aerodynamics, hypersonic vehicle, thermal protection system

1. Introduction

Mars attracted human attention since 1960. To understand the importance of Mars exploration, several reasons should be considered. First, the Red Planet has similar dimensions compared to Earth's continental surfaces. Secondly, it has the capability to host Earth-like ecosystems promoting bacterial life. Furthermore, it represents a prototype sample to study the effects of a probable dramatic climate change [1].

Although the current Martian atmosphere is quite unfeasible for human life, the fundamental component for life i.e., water does exist on Mars around the poles and is probably distributed under the Martian surface. Therefore, a permanent human presence for research purposes on Mars requires reliable and affordable entry vehicles.

Several missions were performed in the past but unfortunately, most of them were unsuccessful [2, 3]. The first attempt to reach Mars was performed by URRS in 1960 with *Marsnik 1*. After several failures just on July 15th, 1965, the NASA *Mariner 4* successfully performed the first Mars flyby. Only in 1976 with *Viking 1* and *2* missions, for the first time, a human device successfully landed on Mars. On 4th July 1997, the first rover, *Pathfinder*, landed on Mars. The purpose of *Pathfinder* was to reduce the costs of space missions. It cost only 5% of *Viking* missions and was developed to test new technologies and approaches for future Mars missions.

During the last decades, significant technological advances have been done. NASA *Pathfinder* successors, *Spirit* and *Opportunity*, provided more than 100,000 images. Currently, three NASA rovers are active on the Martian surface: *InSight, Curiosity*, and *Perseverance*. *InSight* is currently devoted to study the planet's interior; while *Curiosity*, launched in 2012, is still operating on the Gale crater. Finally, Perseverance is the rover of the Mars 2020 mission, and its work objective is to look for signs of ancient life and collect rock and soil samples for a possible return to Earth (i.e., Mars sample return missions) [2]. This mission has the additional record of having allowed a small helicopter, namely Ingenuity, to fly to Mars for the first time in history.

All current Martian devices are unmanned [2]. However, for the upcoming Martian manned missions, recent studies have shown the limited possibility to adopt aeroshells for Mars landing. Braun and Manning [3] highlighted that without a lifting entry, Mars exploration with manned missions can result unfeasible because of the high g-loads associated with hypersonic deceleration and the touch-down.

In this work, a preliminary feasibility analysis of a Martian entry, performed with a lifting body having a blended double delta-wing, is performed. As no previous experience is gained on the use of lifting bodies for Mars entry, a conceptual configuration for an Earth re-entry mission from LEO is preliminarily assumed. A three degree of freedom (dof) trajectory model with a standard Mars atmosphere is adopted to address the entry and descent flight. A simplified heat transfer analysis based on the radiative equilibrium hypothesis for the wall of the descent spacecraft is also performed to compute heat loads during entry. The feedback on a trajectory, obtained by varying AoA and bank angle considering aerodynamic efficiency, and the capability to reach a predefined landing spot is discussed. The possibility to perform a lower deceleration within a shallower entry angle, taking full advantage of the Mars atmosphere is considered.

2. Lifting entry in Martian atmosphere

In 2007 NASA developed a standard design for a reference vehicle architecture for human Mars entry descent and landing (EDL) [4]. It consisted of a hypersonic aero assist entry system with a mid (0.6–0.8) Lift-to-Drag ratio (L/D) aeroshell, able to perform the aerobraking maneuver, that is ejected at a low supersonic Mach number. The use of hypersonic inflatable aerodynamic decelerators (HIADs) is also deeply studied to deliver human-class payloads to the surface of Mars [5]. However, inflatable decelerators do not provide sufficiently low terminal speed adequate for a safe manned landing [6, 7].

Recent studies highlighted the necessity to adopt a different vehicle design for manned missions [2]. A decisive design criterion is represented by the altitude at which the vehicle reaches a subsonic speed. Therefore, lifting body architectures having higher aerodynamic efficiency compared to aeroshells, are currently investigated for a high lift EDL mission [8, 9].

Specifically, lifting bodies (like Dream Chaser) allows a more favorable aerothermal environment and can customize the landing spot using bank angle modulation [10]. Control authority is also important for Mars explorations because it opens the possibility of using a lightweight Thermal Protection System (TPS). Besides, heat load can be controlled with drag (i.e., AoA) modulation which varies the ballistic coefficient during the entry phase according to a prescribed guidance law. Therefore, unlike capsules, a lifting body can take advantage of the

Mars atmosphere, albeit a very thin one, thus reducing the peak heat rate and g-loads.

Although the above-mentioned aspects seem very promising, several issues have to be considered. First, Mars atmosphere is very different from the Earth's one, and past and present lifting body "know-how" is only related to Earth missions. Furthermore, current vehicles prototypes have not performed a complete re-entry from space yet, not even on Earth. Additionally, because of superior aerodynamic performances, the lifting vehicle would tend to skip out because of the thinner Martian atmosphere. Therefore, to perform a conventional entry and descent maneuver, a very shallow entry angle needs to be adopted [4]. Low flight-path angles are preferred in order to achieve a lower terminal velocity to ensure a safe landing phase.

3. Concept vehicle for Mars entry, descent, and landing

To date, lifting body architectures have not been employed yet for Mars EDL missions. Therefore, only *a-priori* knowledge based on Earth re-entry vehicles can be assumed. In this work, a conceptual lifting body configuration, suitable for a manned entry mission starting from a Mars low orbit is considered [8, 9]. The vehicle is derived from a design procedure supported by a multidisciplinary optimization (MDO) analysis. By accounting in the design loop several sub-disciplines, a blended wing body with a double-delta planform configuration with low wing loading, and capable to perform a long gliding trajectory (i.e., $\gamma \ll 1^{\circ}$) can be derived (**Figure 1**) [9, 10].

The vehicle is designed to exploit a specific guidance law modulating the AoA versus Mach number in order to attain a favorable value of convective heat flux profile and promote the cooling down of the passive TPS exclusively by thermal radiation.

Reference geometric parameters of vehicle are shown in **Table 1**.

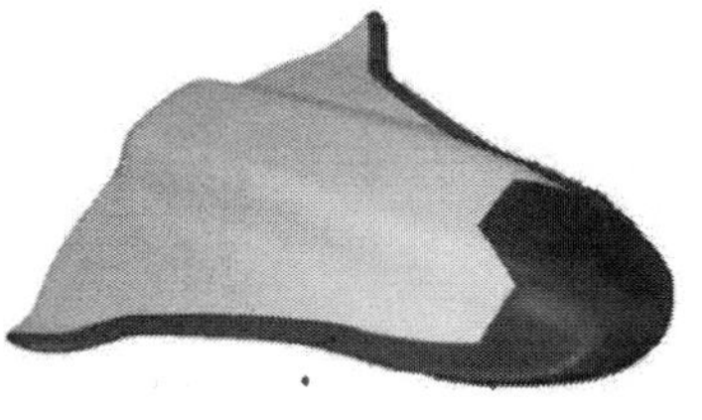

Figure 1.
Entry vehicle representation.

Parameter	Value
Nose radius (R_n)	0.47 m
Reference surface (S_{ref})	44.6 m^2
Mass (m)	12,105 kg
Wall emissivity (ε)	0.8
BC	452 kg/m^2

Table 1.
Entry vehicle parameters.

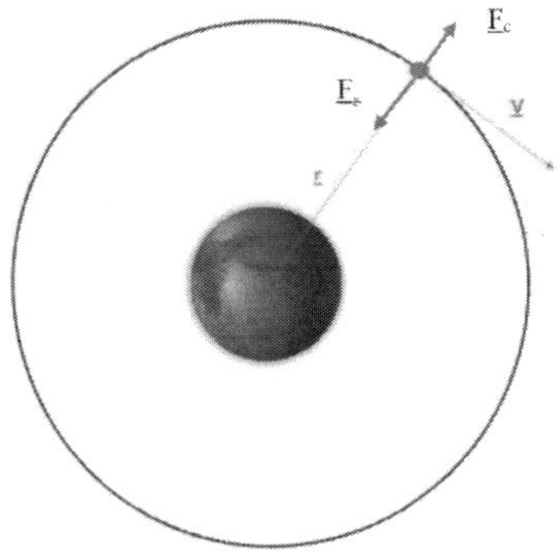

Figure 2.
Initial conditions of entry trajectory.

4. Overall mission specification

The Mars entry flight is described assuming a point mass model, starting from a low circular orbit. The orbit condition implies the equilibrium between centrifugal and gravity force, see **Figure 2**, given by $F_c = \frac{mv^2}{r}$ and $F_g = G\frac{M_{Mars}m}{r^2}$:

$$G\frac{M_{Mars}m}{r^2} = \frac{mv^2}{r} \tag{1}$$

where $r = R_{Mars} + h_0$ and $G = 6.673 \times 10^{-11} \frac{Nm^2}{kg}$, being R_{Mars} the planet radius and h_0 = 120 km the entry altitude.

Using the geometric data for the vehicle, the initial entry speed from Eq. (1) can be obtained:

$$V(t_0) = \sqrt{\frac{G \bullet M_{Mars}}{R_{Mars} + h_0}} = 3484 m/s \tag{2}$$

Initial flight path angle γ_0 is an important design parameter for entry vehicles. For the sake of simplicity, the other initial conditions are chosen as follows:

$$\chi(t_0) = 0\ rad; \theta(t_0) = 0\ rad \varphi; (t_0) = 0\ rad \tag{3}$$

It is supposed that below $M_\infty = 2$ a supersonic inflatable decelerator performs the final deceleration to the terminal landing speed and a paraglider allows the touchdown.

5. Martian atmosphere model

To evaluate the descent trajectory of the lifting body, a proper model of the Martian atmosphere is adopted. In the present computations, the General Circulation Model (GCM) is assumed. GCM provides Mars atmosphere properties according to average data available by past exploration missions on the Red Planet. Atmospheric values of temperature, pressure, and density up to 125 km of altitude (see **Figure 3**) are obtained with the Mars Climate Database (MCD) which allows also small-scale meteorology predictions of storms and seasonal winds [11].

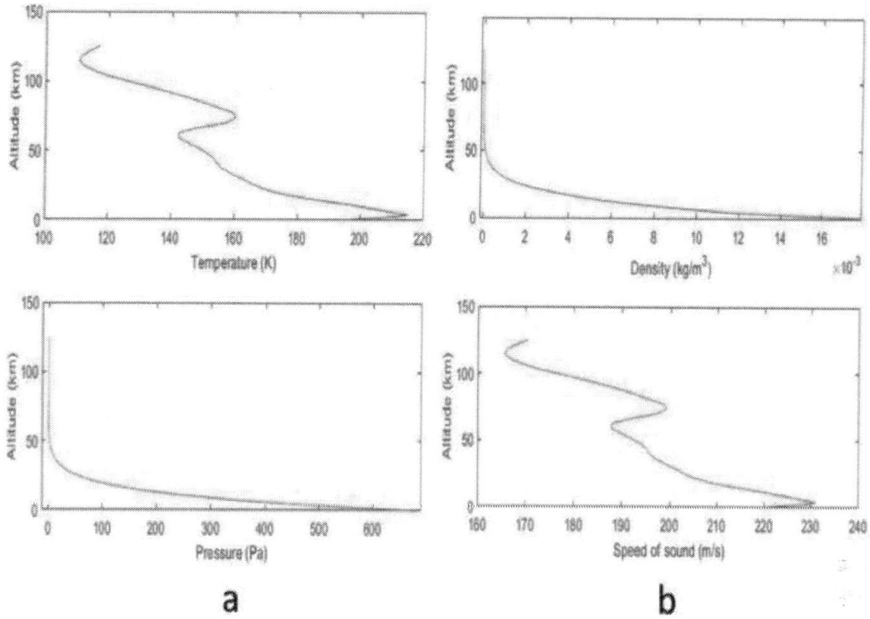

Figure 3.
Martian atmosphere: temperature and pressure (a); density and speed of sound (b).

6. Aeroheating

During most of the entry phase, the vehicle flies at hypersonic speeds. For instance, the aeroshell which landed the NASA rover *Curiosity*, started its descent on Mars at 125 km of altitude, at speed of 5800 m/s [12]. Using the MCD model at 125 km, a $M_\infty = 34$ is obtained. At such high Mach numbers, shock waves and turbulent boundary layers determine very high aerothermal loading conditions (i.e., large radiative and convective heat fluxes, and pressure loads) for the vehicle.

As it is well known, vehicle aeroheating is strictly related to the kind of entry trajectory. Shallower entry leads to a smaller heat rate, but increases the integrated heat loads, thus requiring a thicker TPS. Conversely, steep entries determinate higher heat rates and lower total heat loads. Therefore, in order to adopt a light-weight (passive) and fully reusable TPS, AoA and the flight-path angle are key parameters to trade-off thermal heating and vehicle mass (terminal landing speed).

Generally speaking, the first law of Thermodynamics states that during the descent, the huge amount of kinetic energy (KE) and potential energy (PE) of an entry vehicle dissipate into heat energy that warm up both spacecraft and the atmosphere surrounding it.

Moreover, heat transfer is mutually exchanged between the vehicle surface and the surrounding flow through convection and radiation. The energy balance at spacecraft wall suggests that the convective heat flux, $\dot{q}_{conv}$, is given by the sum of heat conducted into the TPS material, $\dot{q}_{cond}$, minus the amount of heat reradiated, $\dot{q}_{rad}$. By neglecting the conduction inside the heatshield, we have:

$$\dot{q}_{conv} = \dot{q}_{rad} \tag{4}$$

being $\dot{q}_{conv} = 7.207x10^4 \frac{\rho_\infty^{0.47}}{R_n^{0.54}} v^{3.5}$ (W/m^2), given by the Sutton-Graves stagnation-point relationship, and $\dot{q}_{rad} = \sigma\varepsilon T_w^4$, according to the

Stephan-Boltzmann law. Thus, for a safe landing, this huge heat energy must be transferred to the spacecraft shock-layer (i.e., surrounding gas instead of spacecraft) as much as possible.

The heat that goes to the vehicle during time dt is given by the product of heat flux rate, $\dot{q}$, and the reference aerodynamic surface of the vehicle:

$$\frac{dQ}{dt} = \dot{q} S_{ref} \tag{5}$$

which can be re-written as:

$$\dot{q} = \overline{St} \rho_\infty v_\infty \left[\frac{v_\infty^2}{2} + c_p \left(T_\infty - \overline{T}_w \right) \right] \tag{6}$$

where $\overline{St}$ is a mean Stanton number, $\overline{T}_w$ is a vehicle mean temperature and T_∞ is the freestream temperature. Neglecting the thermal contribution $c_p(T_\infty - \overline{T}_w)$, and substituting in (5) one has:

$$dQ = \overline{St} \rho_\infty \frac{{v_\infty}^3}{2} S_{ref} dt \tag{7}$$

Vehicle acceleration related to ballistic coefficient reads:

$$m \frac{dv}{dt} = -\frac{1}{2} \rho v^2 c_D \, S_{ref} \tag{8}$$

Therefore, by replacing S_{ref} given by Eq. (8) in Eq. (7) and integrating over time between initial speed, v_i, and the current one at time t, v, it follows:

$$\Delta Q = \frac{\overline{St}}{2C_D} m \left(v_i^2 - v^2 \right) \tag{9}$$

Eq. (9) expresses the integrated heat load absorbed by the vehicle during the descent in the time interval $[t_i, t]$. Here, spacecraft energy is represented only by KE. Anyway, this approximation can be accepted considering that the PE of entry vehicle is negligible if compared with KE.

According to Eq. (9), the fraction of spacecraft energy that converts into vehicle heating (i.e., ΔQ) depends on the $\overline{St}/c_D$ ratio. Therefore, entry flights with high drag aeroshape (e.g., capsule aeroshapes) are suggested. This design solution, however, is not suitable for manned missions.

High lift configurations must be exploited in order to limit inertial loads and aerothermal loads by flying as much as possible shallow trajectory at higher altitudes, i.e. lower density. However, shallow entries ($\gamma \ll 1$ and $\alpha < 30^\circ$) are characterized by large flight time and consequently large integrated heat loads (ΔQ).

7. Entry dynamics

Mars entry trajectory is computed in a non-rotating, inertial, Mars-Fixed Mars-Centered reference frame [13]. A three degree-of-freedom numeric simulation is assumed, and the spacecraft is described as a point mass, performing a non-planar unpowered descent trajectory with a constant bank angle (**Figure 4**) [13]:

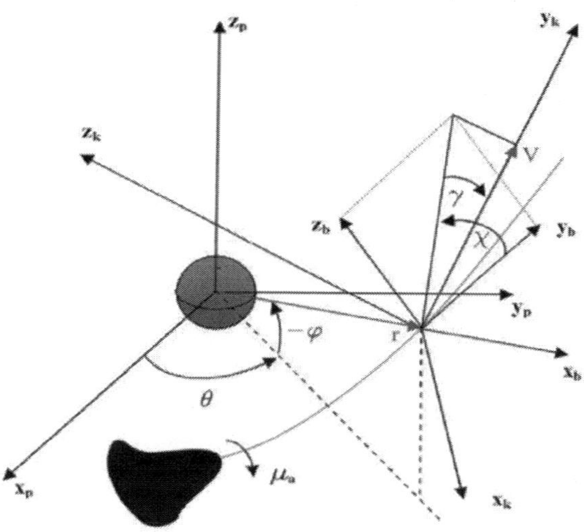

Figure 4.
Reference frames for entry flight equations.

$$\begin{aligned}
\frac{dV}{dt} &= -\frac{D}{m} - g\sin\gamma \\
\frac{d\gamma}{dt} &= \frac{1}{V}\left(\frac{L}{m}\cos\mu_a - g\cos\gamma + \frac{V^2}{r}\cos\gamma\right) \\
\frac{d\chi}{dt} &= \frac{1}{V}\left(\frac{L}{m}\frac{\sin\mu_a}{\cos\gamma} - \frac{V^2}{r}\cos\gamma\cos\chi\tan\varphi\right) \\
\frac{dr}{dt} &= V\sin\gamma \\
\frac{d\theta}{dt} &= \frac{V\cos\gamma\cos\chi}{r\cos\varphi} \\
\frac{d\varphi}{dt} &= \frac{V\cos\gamma\sin\chi}{r}
\end{aligned} \tag{10}$$

where θ, φ are the longitude and latitude in a spherical frame of reference, χ is the heading angle and μ_a is the bank angle.

Lift and Drag force are given by $L = ½\rho V^2 S_{ref} C_L$ and $D = ½\rho V^2 S_{ref} C_D$ respectively where aerodynamic coefficients C_L and C_D are taken using an aerodynamic database in hypersonic and low supersonic regimes [10].

The first order ODE system is integrated over time with a fourth-order explicit Runge–Kutta method.

8. Mars entry and descent flight

8.1 Entry with no bank modulation

Mars lifting entry is studied assuming the initial conditions expressed by Eq. (3):

$$V_e = 3484\,m/s; \gamma_e = -0.3^\circ; \mu_a = 0^\circ; \alpha_e = 20^\circ; 30^\circ; 45^\circ \tag{11}$$

Limits	Values
q_{max}	2000 Pa
n_{max}	0.5 terrestrial g's
Q_{max}	100 kW/m^2

Table 2.
Loads constraints.

In **Table 2** the structural and aerothermal constraints which define the entry corridor are shown.

The descent analysis is first performed at zero bank-angle to compute the nominal vehicle trajectory. With $\mu_a = 0^\circ$ the downrange distance depends only on aerodynamic efficiency (i.e., AoA modulation). In **Figure 5a** the effect of AoA on the entry trajectory is shown.

As α increases, the drag coefficient rises and the descent trajectory sinks more, thus moving to lower altitudes, see **Figure 5b**.

Looking at **Figure 5b** we see that with α = 20° the vehicle reaches M_∞ = 2 at an altitude of about 10 km higher than the case at α = 45°. At α = 20°, the spacecraft features a higher hypersonic aerodynamic efficiency. Therefore, higher L/D values appear more convenient for Mars missions to limit the terminal landing speed allowing at the same time a greater total range.

Figure 5c-5d confirms the advantage of flying with a high L/D. Long entry time (one-hour order-of-magnitude) could also help with the landing spot customization. However, for lifting bodies, increasing L/D leads to an increase of flight time (see **Figure 5d**), and integrated heat loads, see Eq. (5).

Finally, **Figure 5e-5f** show the effect of α on the peak heat flux rate and on its time history. At $\alpha = 45^\circ$ the peak heat flux is 90 kW/m^2, while decreases to 50 kW/m^2 at $\alpha = 20^\circ$. Therefore, the thermal peak is reduced, and non-ablating (re-usable) materials can be adopted. On the other hand, the longer flight time requires high emissivity materials to decrease heat transfer by conduction. In **Table 3** it is shown the total energy absorbed during entry.

In **Figure 5c** it is shown that for $\gamma = -0.3^\circ$ entry times are of the order of 10^4 s. To reduce entry flight time, steeper flight-path angles can be considered, see **Figure 6**.

Assuming $\gamma = -2.4°$, entry times are of the order of 10^3 s which can be compared to Earth re-entry flight time. However, progressively decreasing γ the peak heat flux rises (see values attained at $\gamma = -2.4°$). Therefore, a trade-off study between entry time and flight path-angle suggests the appropriate value related also to thermal insulation material capabilities.

8.2 Entry with bank angle modulation

Aerodynamic efficiency can be varied implicitly without using AoA modulation opportunely changing the bank angle μ_a. Bank modulates the lift force into two components $L_{eff} = L \cos\mu_a$ and $L_{side} = L \sin\mu_a$. The lateral force, L_{side}, deviates the lift vector outside the plane of trajectory, thus allowing a cross-range performance in the lateral direction. Therefore, trajectory footprint can be decomposed into two components: the down-range $\Delta x_p \sim \frac{L}{D} \cos\mu_a$ and the cross-range $\Delta y_p \sim \frac{L}{D} \sin\mu_a$. In **Figure 7** the effect bank modulation on vehicle trajectory is shown.

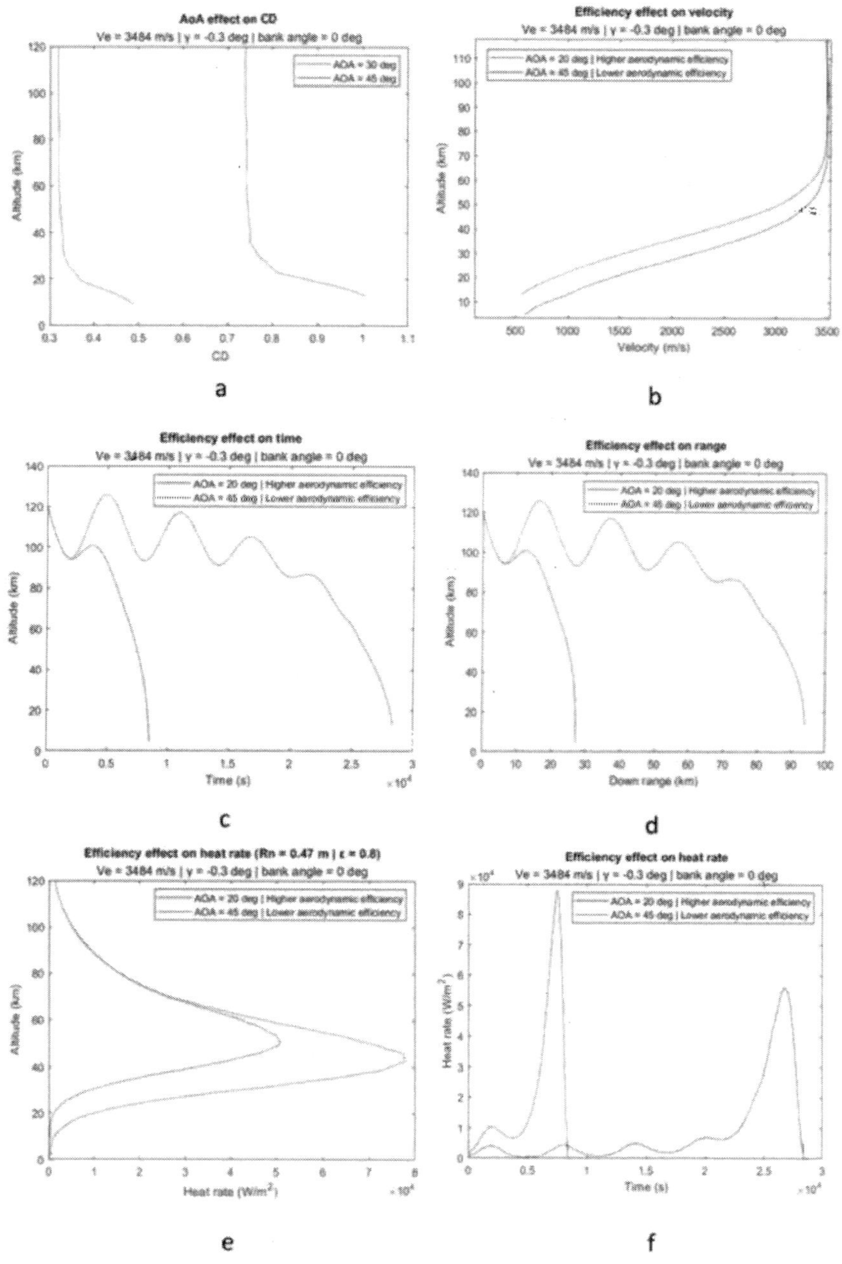

Figure 5.
Entry analysis with no bank angle modulation.

AoA (deg)	Total heat load (MJ/m^2)
20	374.72
45	113

Table 3.
Total energy absorbed during entry.

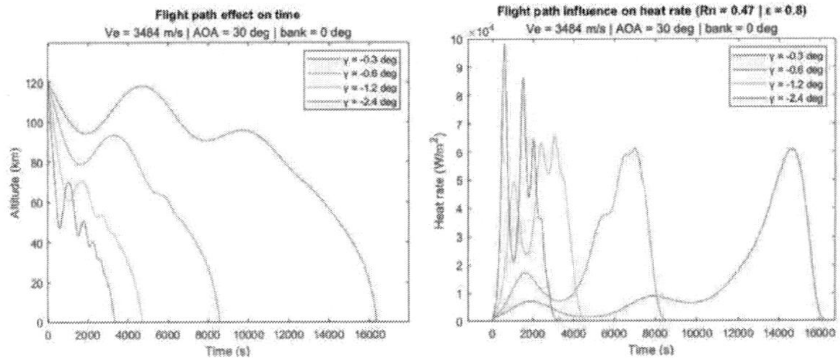

Figure 6.
Trade-off effects of flight path angle for entry flight at $v_e = 3484$ m/s, $\alpha = 30°$, $\mu_a = 0°$.

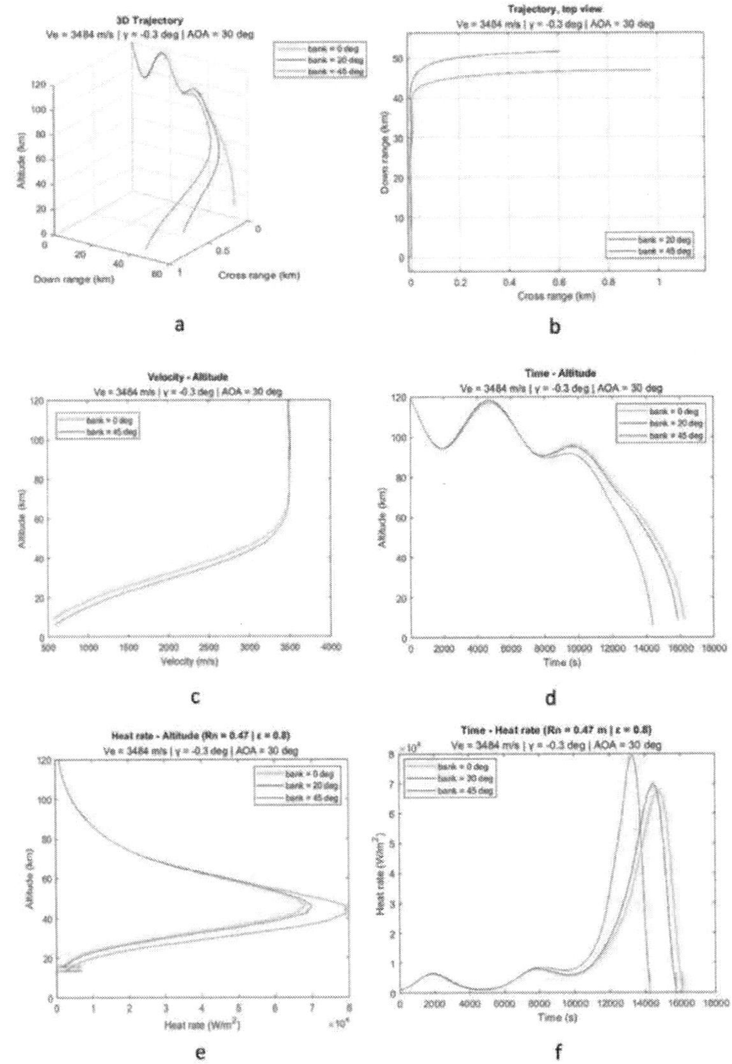

Figure 7.
Entry trade-off analysis with bank angle modulation.

Bank angle (deg)	Total heat load (MJ/m^2)
0	226.34
20	220.95
45	198.88

Table 4.
Total energy absorbed during re-entry.

Increasing μ_a at constant AoA (i.e., constant L/D) the total range (cross-range and downrange) increases. The effective lift component L_{eff} is reduced, and the vehicle attains a lower altitude at $M_\infty = 2, \mu_a = 20^\circ$ with respect to $\mu_a = 40^\circ$ as it is shown in **Figure 5c**. Therefore, control authority increases at expense of a higher terminal speed. Another effect of bank angle modulation can be observed in **Figure 7d**. The lower lift reduces the skip motion when the vehicle encounters the Martian atmosphere. Higher heat flux peaks over a shorter entry time are experienced. Therefore, at constant AoA, bank modulation improves the capability to reach the predefined landing spot.

Finally, in **Table 4** the total energy absorbed during the entry with bank modulation is computed.

From an energetic point of view, bank modulation toward higher μ_a favors lower total energy stored but shifts trajectory toward a ballistic-like behavior.

9. Conclusions

In this work, the feasibility of Mars entry with a lifting body was studied. Aerodynamic efficiency control is studied with respect to AoA and bank angle modulation. Decreasing AoA during Mars entry allows to reduce the peak heating, consequently increasing the flight-time. However, higher control on trajectory can be performed with bank angle modulation, which reflects over a higher control on the landing point, as well as the possibility to have a cross range. Furthermore, higher efficiency allows the vehicle to decelerate more at a higher altitude, taking advantage of the thin atmosphere. This is a crucial aspect for Mars missions in order to limit the terminal speed within allowable values for manned missions. Deceleration allows more tolerance on parachute deployment. Another benefit of a lifting vehicle is the influence on heat rate and total heat load: by increasing the aerodynamic efficiency, the heat rate peak can be decreased. The longer flight time induces the vehicle to store a higher amount of energy from the atmosphere, increasing the total heat load. Therefore, bank modulation can be performed assuming a suitable guidance law that modulates the vehicle aero-thermal loads inside the predefined constraints.

References

[1] Genova A. Oracle: A mission concept to study Mars' climate, surface and interior. Acta Astronautica. 2020;**166**: 317-329

[2] NASA. Mars 2020 Perseverance Rover. 2020. Available from: https://www.mars.nasa.gov/programmissions/missions/future/mars2020

[3] Braun RD, Manning RM. Mars exploration entry, descent and landing challenges. Journal of Spacecraft and Rockets. 2007;**44**(2):310-323

[4] Mall K, Michael JG. High mass Mars exploration using slender entry vehicles. In: AIAA Atmospheric Flight Mechanics Conference. Reston, VA: American Institute of Aeronautics and Astronautics; 2016. DOI: 10.2514/6.2016-0019. Available from: https://arc.aiaa.org/doi/abs/10.2514/6.2016-0019

[5] Sostaric R. The Challenge of Mars EDL (Entry, Descent, and Landing). NASA JSC-CN-20470. Houston, TX, United States: NASA Johnson Space Center; Apr. 2010

[6] Lu Y. Aerocapture, aerobraking, and entry for robotic and human Mars mission. In: Pezzella G, Viviani A, editors. Mars Exploration A Step Forward. IntechOpen; 2020. DOI: 10.5772/intechopen.93281

[7] Viviani A, Iuspa L, Aprovitola A. Multi-objective optimization for re-entry spacecraft conceptual design using a free-form shape generator. Aerospace Science and Technology. 2017;**71**:312-324

[8] Lafleur JM, Cerimele CJ. Angle of attack modulation for Mars entry terminal state optimization. In: AIAA Atmospheric Flight Mechanics Conference; 10–13 August 2009. Chicago. Aerospace Research Central. Reston, VA: American Institute of Aeronautics and Astronautics; 2009

[9] Aprovitola A, Montella N, Iuspa L, Pezzella G. An optimal heat-flux targeting procedure for LEO re-entry of reusable vehicles. Aerospace Science and Technology. 2021;**112**:106608

[10] Viviani A, Aprovitola A, Iuspa L, Pezzella G. Aeroshape design of reusable re-entry vehicles by multidisciplinary optimization and computational fluid dynamics. Aerospace Science and Technology. 2020;**105**:106029

[11] Mars Climate Database (Web interface). 2016. Available from: http://www-mars.lmd.jussieu.fr/

[12] NASA. Final Minutes of Curiosity's Arrival at Mars. 2010. Available from: https://www.nasa.gov/mission_pages/msl/multimedia/gallery/pia13282.html

[13] Hirschel EH, Weiland C. Selected Aerothermodynamic Design Problem of Hypersonic Flight Vehicles. Berlin Heidelberg: Springer; 2009

Index